Alexander & His Adversaries

By Will Nesbitt

Credits

Written by / Game Design by Will Nesbitt

Edited by Richard Evers

Art by Will Nesbitt

Layout by Gilbert Rafanan

Miniature Photography and Painting By

New Buckenham Historical Wargamers

Kevin Krause

Richard Evers

JC Lira

Alexey Kalinin, Art of Wargame painting service

Foreword

Your Guide

Aristotle and Calonos are two of the better known philosophers who advised Alexander of Macedonia. I am one of Alexander's lesser known tutors and I am called the D10 Philosopher. I will be your guide in the reference and will from time to time add commentary or further information which I think may be helpful to understanding the rules or the era.

I'll start by telling you that this is a sourcebook supporting the core rules of the *Blood of Ancients* system published by Military Miniature. *Blood of Ancients* is a generic rules-system for using 28 mm figures on a 110 mm hex grid. *Blood of Ancients* has detailed rules about the charge and counter-charge system, but it really doesn't have much in the way of specifics for armies.

Alexander & His Adversaries focuses on Alexander the Great and specifics about armies that fought during the era of Alexander. Herein you will find army lists for various forces with new rules specific to armies of this era. *Alexander & His Adversaries* provides army lists and set-up for a number of important and famous battles from the era. In addition, the appendices explain how to play *Blood of Ancients* without a hex grid and how to convert this game to any scale so that hobbyists can use the figures they already have.

Alexander & His Adversaries presumes an understanding of the *Blood of Ancients* core rules. In the following preface, you will find a summary of core rules. These are not a substitute for the actual rules, but they can give you a good idea about the foundation this reference stands upon.

Image courtesy of Richard Evers.

Image courtesy of Richard Evers.

Preface

Alexander & His Adversaries is a sourcebook for and supplement to *Blood of Ancients*. *Blood of Ancients* is not required to use this book, but a basic understanding of the game's framework will be useful for implementing the rules presented here. For this reason, **a summary of the *Blood of Ancients* core rules is provided in the first Appendix of these rules**.

Herein we also present set-ups and scenario notes to play out numerous battles from Alexander's march to glory. This sourcebook also contains army lists and troop profiles relevant to the age of Alexander. But before we delve into any of that, this preface contains a quick overview on how to read army lists and Leader profiles.

> ### No Hex Grid Required
> The appendices in this sourcebook will give an overview of how to play this game without a hex grid and how to play with miniatures of any scale. This overview is abbreviated from the full explanation found in *Rome & Her Rivals*, another *Blood of Ancients* companion.

Image courtesy of Richard Evers.

Army Lists

The Units profiles in *Alexander & His Adversaries* are suggestions of how you might want to field models that reflect the described armies. The full definitions the terms in these profiles values can be found in *Blood of Ancients*, but as a quick reminder:

Strength

The number of dice a unit rolls in attack and the number of casualties a unit cannot suffer. Also the base Discipline of the unit.

Discipline

This is a bonus or penalty to Morale Checks and Courage rolls.

Vulnerability

An attacker needs to roll this number or lower to score a hit on the unit. Do not add this to the base defense in the *Blood of Ancients* Quick Reference. In fact, this is just a precalculated version of the number that you should arrive at when looking at the Quick Reference.

Whiff

Whiff represents a combatant's tendency to err in combat. Add Whiff to a combatant's die roll, as the goal is always to roll low. A negative Whiff subtracts from the attacker's roll, and is thus a benefit.

> ### Regarding Unit Profiles
>
> In a 1:1 Ratio of Troops to Strength, Strength is the number of models in the Unit. You can read more about how to customize these rules to fit the scale of models in your collection in the appendix and in the *Blood of Ancients* sourcebook *Rome & Her Rivals*.
>
> As you peruse the profiles of various armies of the error, keep in mind that *Blood of Ancients* is a game first and a simulation second. Therefore, some of the profiles will not match exactly with their historical counterparts. Finally, **except in tournament play**, if you don't find a troop profile for a particular army, you can substitute from another army list to get an idea of the value of the troop.

Image courtesy of Kevin Krause.

Alexander & His Adversaries

Table of Contents

Foreword	iii
Preface	v
Leaders	1
Alexander's Upbringing	13
Alexander's Armies	19
Alexander Becomes King	37
Alexander Invades Persia	49
Achaemenid Persia	59
Founding of Alexandria	69
Eastern Persia	71
Bactria and Sogdiana	77
To the Ends of the World	79
India	83
Era-Specific Rules	89
Appendix I: *Blood of Ancients*	95
Appendix II: Personalizing Fit To Your Collection	99
Appendix III: Playing Without A Hex Grid	103

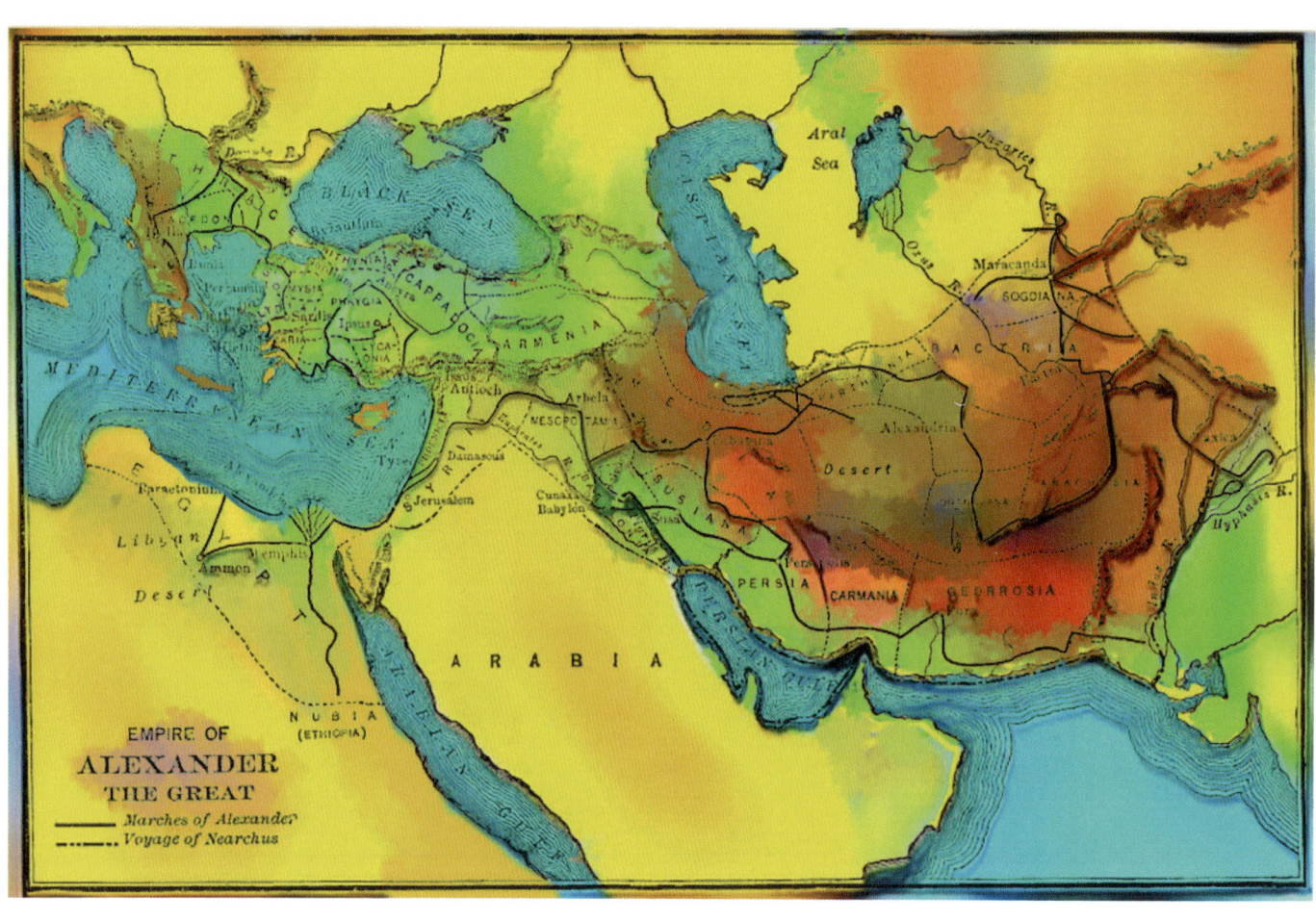

Leaders

Alexander the Great is considered one of the greatest battlefield leaders in history. His military genius and tactical innovations led to numerous victories and the creation of one of the largest empires in the ancient world. He was known for leading from the front and fighting alongside his men, which boosted their morale and made him a respected and inspirational figure.

Thus, it all starts with leadership. In antiquity, leaders played a crucial role in shaping the outcome of battles. This sourcebook introduces a number of advanced rules to recreate the ability of these leaders to overcome their opponents by sheer force of will, superior martial skill, and innovative tactics.

In armies of antiquity, it was not uncommon for commanders to lead from the front and participate in the battles themselves. Philip II of Macedon, the father of Alexander the Great, lost his right eye during the siege of Methone in 354 BC. During this siege, he was hit in the eye by an arrow. Despite this injury, Philip continued to lead his troops effectively and managed to conquer the city.

The practice of leading from the front was seen as a way to inspire troops, demonstrate courage, and enhance the morale of the soldiers. It also helped to cement the commander's authority and reputation as a brave and skilled warrior.

However, the necessity for a commander to lead from the front and personally fight in battles varied depending on the culture, battlefield circumstances, and individual leader. Some commanders, like Darius III, might not have fought personally in every battle or issued challenges to their enemies. Instead, they might have relied on their tactical and strategic abilities to command and control their forces from a safe distance. In some cases, the approach of laying back and staying above the fray allowed a leader to better assess the situation and make informed decisions, but this approach could also make them seem less connected to their troops.

In these rules, each army is led by a prominent historical figure, representing the general in command of the forces. Leaders are typically associated with a specific faction or kingdom, and their abilities reflect

About the Organization of this Sourcebook

In some ways, this reference is organized as a historical narrative tracing Alexander's career from heir, to king, to emperor and then to the battle for his empire after his untimely death. In each stage of Alexander's story we will introduce a few more rules, an army list, and troop types, providing such rules as are necessary to game out that part of Alexander's career. At each stage of Alexander's career we provide at least one scenario to play.

The rules and armies presented herein are presented in the order in which Alexander encountered them. After Alexander's death, we provide rules for competitive play so that players can organize tournaments and campaigns.

Finally, the Appendices provide references that summarize core rules from Blood of Ancients and lay out advance rules that you may wish to add to your game once you understand the basics.

the unique strategies and tactics employed by their respective forces. All Leaders have the following characteristics:

- Leadership,
- Tactical Advantage,
- Inspirations, and
- Limitations.

Leadership

In this sourcebook, each leader has an attribute called Leadership. For example, Alexander the Great's Leadership is greater than the Leadership of Darius III. A leader's Leadership is equal to his Command Radius, which is quite simply the number of hexes his authority extends. Units outside of the leader's Command Radius are out of command. But as you'll see in the rules that follow, Leadership has other uses as well.

Tactical Advantage

Tactical Advantage represents the additional benefit a commander provides to a unit they are attached to. The higher the Tactical Advantage value, the greater the impact the commander has on the unit's performance in battle.

Tactical Advantage equals a number of extra attack dice that are rolled by the unit during combat. This

> ### In Blood of Ancients all Leaders are the same.
>
> In *Blood of Ancients* all leaders have the same Command Radius and the same Inspirations. In *Alexander and his Adversaries*, Leadership is an attribute and the Inspirations that a Leader has will vary from leader to leader. Leadership is the same thing as Command Radius, but has other uses which are described later.

represents the enhanced capabilities and effectiveness of the unit due to the commander's presence, personal strength, and guidance.

Leaders in Melee

In order to use the Tactical Advantage bonus, a leader must attach to a friendly unit. A leader may attach to any friendly unit in his Command Radius during the Command Movement Phase. When a leader joins a unit, the leader is considered to be part of that unit for the duration of the turn combat.

When a leader is part of a unit, the unit gains a bonus to its number of attacks, ranging from -1 to +3, depending on the leader. A negative number of attacks indicates that the unit must expend efforts to protect the Leader in battle. This bonus applies as long as the leader remains with the unit.

Image courtesy of Alexey Kalinin.

Image courtesy of New Buckenham Historical Wargamers.

Targeting Leaders

The enemy cannot specifically target a leader within a unit during combat. Instead, any damage dealt to the unit is considered to be dealt to the unit as a whole. Each time a unit with an attached leader takes a Disruption from melee or missile fire, the leader rolls a die. On a roll of a ten the leader has been routed or injured. Immediately roll again. If a ten is rolled again, the leader has perished. Rules for what happens when a leader is routed, injured or perished are found below.

Alexander In Battle.

During the Battle of Issus in 333 BC, Darius III, ruler of the Persian Empire, was at the center of the Persian line surrounded by personal bodyguards. Alexander demonstrated his personal courage and skill as a warrior by leading a cavalry charge directly at Darius. Alexander and his elite Companion Cavalry broke through the enemy lines, causing chaos and confusion among the Persian forces.

As Alexander approached Darius, he found himself in close combat with a group of Persian nobles who were trying to protect their king. In the melee, Alexander was said to have killed several of these nobles, including one of Darius' own relatives. Faced with the prospect of a personal confrontation with Alexander, Darius decided to flee the battlefield, leaving his army in disarray and ultimately leading to the Persian defeat.

This anecdote not only showcases Alexander's fearlessness in battle, but also highlights his determination to take on the most dangerous tasks himself. His willingness to charge directly at the enemy king and engage in hand-to-hand combat with his elite guards inspired his troops and contributed to his reputation as a great military leader.

Image courtesy of New Buckenham Historical Wargamers.

Leader Rout

If a unit with a leader fails a Morale check and routs, the leader must make a separate Morale check. To pass the check the player must roll equal to or less than the leader's Leadership with a +3 bonus to its Leadership attribute. If the leader passes the Morale check, it can move to an adjacent friendly unit that is within its Command Radius. If the leader fails the Morale check, it routs with the unit.

Leader Recovery

A wounded leader can recover during the Reinforcement phase. Roll a die at the beginning of the Recovery Phase; if the result is equal to or lower than Leadership, the leader recovers from its wounds and regains his bonuses for the remainder of the game. Because of the Rule of Aces, a roll of a 1 means that the leader fully recovers. When a leader recovers from his injuries, he must convalesce for 1 turn in a unit that is not adjacent to the enemy or in Melee combat. Furthermore, a leader may not inspire while convalescing.

Leader Replacement

If a leader is killed, the player can replace the leader with another leader from their army reserve during Recover. The replacement leader has a Command Radius (Leadership) and Value that is 1 less than the original leader. The replacement leader has no special Inspirations.

Inspirations

All leaders have access to a set of basic Inspirations, which represent general tactical and strategic abilities that any skilled Leader might employ on the battlefield. An Inspiration is a leadership action. Only Units that are within a leader's Command Radius can benefit from Inspiration. The standard Inspirations described in *Blood of Ancients* and repeated herein can be used by any leader, regardless of faction or kingdom, and are available in addition to their special Inspirations.

The special Inspirations represent a leader's unique ability to motivate, direct, and inspire their troops to perform in ways that other leaders cannot. These are mentioned in the leader's profile and described fully in the Leader's chapter of this sourcebook.

A leader can inspire once per turn every single turn. An Inspiration can be declared at any time, even out of Phase or when a player is not active, but the effects occur during the appropriate phase of the turn, as outlined in the game rules. Any leader can apply any of the standard Inspiration. But this reference introduces some special Inspirations which a leader can only use if it's on the leader's profile.

Inspirations may not "stack" or combine their effects unless the description specifically states otherwise. This means that a unit cannot benefit from multiple Inspirations of the same or different types in a single turn, unless explicitly allowed by the Inspiration's description.

When looking at the benefits of Inspirations, it's important to remember that +1 bonus to Attack is the same thing as a -1 to Whiff. By the same token, a +1 to Defense is a -1 to Vulnerability.

Standard Inspirations

These are available to all leaders.

Seize the Initiative

A player may use a leader to seize or force initiative. If both players Seize the Initiative, roll a die and add Command to determine the winner. (Declare at the beginning of the turn.)

Push

Push grants one extra point of Movement for any Unit in command. (Declare during Movement.)

Furious Attack

When a leader inspires a Furious Attack, the attacking unit immediately rerolls all Attack dice. (Declare after rolling Attack Dice.)

Stalwart Defense

When a leader inspires a Stalwart Defense, the attacker must reroll all Attack dice. (Declare after the defender sees the attacker's result.)

Image courtesy of JC Lira.

Rally

When a leader inspires a Unit to Rally, the Unit may reroll a Morale or Courage Check. (Declare immediately after a failed roll.)

Special Inspirations

Special Inspirations are only available if a leader in play has a particular special Inspiration on his profile. A leader may only use one Inspiration per turn whether that Inspiration is a special Inspiration or a standard Inspiration.

Alexander's Charge

Alexander's Charge is a testament to the legendary Macedonian prince's ferocity and skill on the battlefield. With this Inspiration, the attached cavalry unit channels the spirit of Alexander himself, striking fear into the hearts of their enemies. As the unit charges, their momentum surges, gaining an additional +2 to +4 charge bonus. This powerful and awe-inspiring attack has the potential to turn the tide of battle in a single, decisive moment.

Disciplinarian

When this Inspiration is used, the commander can select an out-of-command unit, regardless of the command range. For the duration of the turn, the selected unit functions as if it is in command, allowing it to perform actions that would otherwise be restricted. This exception to the standard rule of Inspirations allows the commander to exert their influence over units that are out of command, ensuring that they remain effective on the battlefield.

Dependable Support

When this Inspiration is used, the commander confers a bonus to a friendly unit engaged in melee based on the number of supporting units. A supporting unit is defined as up to 2 friendly units adjacent to the rear of the attacking unit. The commander must be in one of the supporting units. The melee bonus is +1 for each supporting unit, up to a maximum of +2. The player must declare the use of this Inspiration and choose the attacking unit to receive the bonus before resolving the melee combat.

Fierce

Fierce embodies the relentless aggression and determination of history's most tenacious warriors. When a unit is inspired by this potent ability, they unleash their full fury upon the enemy, gaining a +2 bonus to their attacks. The overwhelming force of these warriors can break through enemy lines, carving a path of destruction and leaving opponents reeling in their wake. This Inspiration can make the difference between victory and defeat, as the sheer intensity of the attack catches enemies off guard.

Guerilla

Inspired unit can move through Rough Terrain as if it were Open Terrain, allowing the unit to navigate difficult landscapes with ease and surprise their enemies.

After a Melee or Ranged attack, when this leader is attached, the unit may move 1 hex in without regard for facing or Maneuver costs, reforming with any facing. This means that after the attack, but before the Morale roles, a Guerilla may disengage and reposition themselves, avoiding retaliation from the enemy.

Innovator of the Phalanx

Innovator of the Phalanx represents Philip's mastery of advanced tactical formations and innovative battlefield maneuvers. The inspired unit gains the ability to perform one free maneuver, showcasing unparalleled battlefield coordination and discipline. This rare tactical acumen enables Philip's army to adapt and respond to enemy movements with ease, often turning the tide of battle in their favor. Through the creative application of this Inspiration, a player can exploit weaknesses and seize opportunities on the battlefield.

Inspiring Presence

When this Inspiration is used, the commander can target a friendly unit that is either attached or adjacent to the leader. For the duration of the turn, the targeted unit gains a +2 morale bonus, making it more resistant to morale checks and less likely to waver in the face of adversity. This represents the commander's ability to rally and encourage their troops during critical moments in battle when they are in close proximity.

Loyal Friend

When this Inspiration is used, the commander can target another friendly commander who is either attached or adjacent to the leader. For the duration of the turn, the targeted commander's Inspirations gain an additional +1 bonus to their effects, signifying the strong bond and synergy between the two commanders. This represents the power of friendship and loyalty in enhancing the effectiveness of their leadership on the battlefield. Loyal Friend is an exception to the stacking rule and can be combined with other Inspirations, as long as it targets another friendly commander.

Master of Maneuver

When this Inspiration is used, the inspired unit is granted up to 2 free Move points that must be used exclusively for maneuvering. This means that an infantry unit, for example, can move 3 hexes and also perform up to 2 maneuvers during the move. That same unit could not use this inspiration to move 5 hexes straight ahead. Master of Maneuver provides units with additional tactical flexibility, enabling them to better adapt to changing battlefield conditions or exploit openings in the enemy lines.

Mountain Fighter

When this Inspiration is used, the inspired unit gains a one-time advantage while fighting in mountainous terrain. The unit can ignore the movement penalties associated with mountains and rough ground. Or, the unit gains a +2 bonus to attack for one turn when the contested ground is rough terrain. This represents the commander's expertise in leveraging the unique aspects of mountainous terrain to their advantage, allowing the unit to maneuver more effectively and strike with greater force.

Raider

The inspired unit gains the ability to move or maneuver through an enemy zones of control without stopping. This only affects a single hex.

Reliable Commander

When this Inspiration is used, the inspired unit may reroll a single failed combat result or Morale check during a single turn. This represents the commander's ability to instill confidence and determination in the unit, allowing them to recover quickly from setbacks and maintain their fighting spirit. The player must decide to use this Inspiration immediately after the failed roll and may only reroll once.

Resourceful

When this Inspiration is used during the Recovery phase, the commander's resourcefulness allows the target unit to recover without paying the usual price. Normally, each time a unit recovers, 1 Strength is removed from the unit as a casualty. However, when this Inspiration is applied, the target unit recovers without losing any Strength. The player must declare the use of this Inspiration before applying it to the target unit.

Royal Authority

When this Inspiration is used, the commander's prestigious status as a royal figure grants a one-time bonus to a target unit within command range. The unit receives a +1 modifier to both Attack and Defense values for a single round of combat. The player must declare the use of this Inspiration before the unit engages in combat. The aura of the commander's royal authority bolsters the troops' confidence and fighting spirit, leading to enhanced performance on the battlefield.

Seaborne

This Inspiration represents the commander's experience in coordinating attacks and maneuvers along coastal areas or near bodies of water. When this Inspiration is used, a target unit within the commander's command range gains a +1 bonus to Movement Points and a +1 bonus to Attack or Defense for one turn, as long as they are moving or fighting in a hex adjacent to a river, lake, or coast. This bonus reflects the commander's ability to utilize the surrounding terrain to their advantage, benefiting from their knowledge of naval and coastal warfare tactics.

When using this inspiration, the commander's scouting abilities come into play. For one turn, any enemy unit within the commander's command range suffers a +1 to Vulnerability. This represents the commander's efforts before the battle began. By gathering valuable information about the enemy's movements and plans, friendly units can exploit weaknesses, causing an increase in Vulnerability.

Skirmish Expert

When using this inspiration, the commander's knowledge of skirmish tactics greatly benefits the friendly units. The inspired unit gains a +2 bonus to their attack roll. This represents the commander's ability to direct and coordinate the actions of their skirmishers, making them more effective at harassing and disrupting enemy formations.

Steadfast

Steadfast represents the unwavering resolve and ironclad discipline of a battle-hardened unit under the guidance of an exceptional commander. The inspired unit gains a +1 bonus to defend against Melee attacks, showcasing their superior defensive skills and ability to hold their ground in the face of adversity. This Inspiration instills confidence and determination in the unit, empowering them to repel enemy assaults and stand firm in the chaos of battle. A Steadfast unit can help a commander create an unbreakable defensive line that frustrates and demoralizes the enemy.

Tactically Flexible

When using this inspiration, the commander's adaptability and ability to adjust to changing circumstances on the battlefield come into play. For one turn, any friendly unit within the commander's command range can change their facing or perform one additional maneuver without the usual movement cost. This represents the commander's ability to recognize opportunities and quickly adapt their troops' positioning to exploit them.

Tactician

When using this inspiration, the commander's tactical expertise comes into play. For one turn, any friendly unit within the commander's command range gains a +1 bonus to their combat rolls (both melee and

Image courtesy of New Buckenham Historical Wargamers.

ranged attacks). This represents the commander's ability to analyze the battlefield situation and make quick, effective decisions that give their troops an advantage in combat.

Unbreakable Will

The inspired unit automatically passes a Morale check. Unbreakable Will exemplifies unwavering courage and tenacity of a unit under the influence of a remarkable leader. The commander's presence and guidance bolster the spirits of the soldiers, ensuring they fight to the last many. As a result, the inspired unit automatically passes a Morale check, fearlessly confronting challenges and remaining stalwart come what may.

Veteran Commander

When using this inspiration, the commander's wealth of experience and knowledge of warfare comes into play. One Disruption is removed. In other words the inspired unit regains 1 Strength. This Inspiration will not restore Casualties to play.

Limitations

A Limitation is a constraint or weakness associated with a leader, reflecting their personal flaws, fears, or circumstances that may negatively impact their performance on the battlefield. These factors can be exploited by the opposing player to gain a temporary advantage during the game.

> ### When you are your own worst enemy ...
>
> When a player opts to activate an opponent's Limitation, they are essentially capitalizing on the vulnerabilities of the opposing leader, potentially hindering their effectiveness or causing them to make suboptimal decisions. Limitations can be activated in place of using an Inspiration, adding an additional layer of strategy to the game as players must weigh the benefits of using their own Inspirations against the potential impact of activating an opponent's Limitation. With the rule preventing the same Limitation from being activated on consecutive turns, players are encouraged to vary their tactics and consider multiple options throughout the game.

Each leader can use one Inspiration per turn. Alternatively, the player can forfeit a leader's Inspiration and instead opt to activate an opponent's Limitation. However, a player cannot activate the same Limitation on consecutive turns.

Limitations are weaknesses or tactical difficulties associated with a leader's style of command or their kingdom's historical context. From a game stand point, Limitations play exactly like Inspirations which is to say that a player can declare a Limitation out of sequence at any point that Limitation can affect play. A Limitation can only be played once per turn but it can be played every single turn. The primary difference between Limitations and Inspirations is that a Limitation is played by the opponent.

A player can cancel a Limitation by forfeiting his Inspiration, but only if the leader has not already used his Inspiration this turn.

Age

The commander's age has begun to affect their physical and mental abilities, making them less effective on the battlefield. When this limitation is played by the opponent, the affected leader cannot use his Tactical Advantage for that turn.

Ambitious

The commander is overly ambitious, which can cause them to make rash decisions or overextend their forces in pursuit of personal glory. When this limitation is played by the opponent, the affected commander must spend their next activation attempting to achieve a significant objective, such as capturing an enemy leader, seizing an important strategic location, or defeating a powerful enemy unit. This may result in the commander's forces becoming overextended, leaving them vulnerable to counterattacks. This limitation can only be overridden by using the commander's Inspiration for the turn.

Aspiring Leader

This commander is less effective than the commanding general and can at times be overwhelmed by their circumstances. The opponent picks one unit that is in command, and that unit functions as if it is out of command (dice for results as if the unit was outside of the leader's command radius). The leader can at any time override this limitation, but doing so uses that leader's Inspiration for the turn.

Brash

The commander's overconfidence leads them to take unnecessary risks on the battlefield. When

> ### Example
>
> For example, Philip's Limitation is Fatherly Concern, which is played by the opponent and not by Philip. Philip's Inspiration is the Innovator of the Phalanx, plus all the standard Inspirations.

this limitation is played by the opponent, the affected leader's unit must charge the nearest enemy unit if possible, even if it's a disadvantageous engagement. If no enemy unit is in range, the leader's unit must move towards the nearest enemy unit. This limitation represents the leader's reckless determination to prove their prowess in battle.

The commander is overly cautious, making them less likely to take risks that could lead to significant gains. When this limitation is played by the opponent, the affected commander must prioritize the safety of their own unit and any units under their direct command. For their next activation, they must avoid initiating any high-risk actions such as charging into combat, pursuing retreating enemies, or advancing into potentially dangerous territory. This limitation can only be overridden by using the commander's Inspiration for the turn.

Cowardice

A coward fears death more than dishonor. If a unit adjacent to, or attached to, this leader retreats, then this leader withdraws from the field and is removed from play.

Fatherly Concern

Philip must concern himself with the safety of his son Alexander. When this limitation is played by the opponent, Philip must prioritize the safety of Alexander. Philip must issue an order that Alexander to withdraw from any potentially dangerous situation (i.e. retreat and disengage from melee). This limitation can only be overridden by using Philip or Alexander's Inspiration for the turn. This Limitation is declared during the Movement or Command Movement phase.

Fear of Defeat

The commander is overly cautious due to a fear of losing, which can lead to missed opportunities on the battlefield. When this limitation is played by the opponent, the affected leader's units cannot take advantage of any opportunities to charge the enemy or exploit gaps in the enemy lines during that turn.

Impetuous

This limitation allows the opponent to select a unit adjacent to or attached to the leader. That unit is required to charge an enemy unit, pursue a retreating enemy, or engage in close combat.

Local Loyalties

The commander has strong ties to the local population, which can sometimes conflict with their duty to the overall campaign. When this limitation is played by the opponent, the affected leader must prioritize the safety and well-being of a unit in his command. The unit, selected by the opponent, may not charge or engage in offensive actions.

Loyalties

The commander is extremely loyal to the overall commander, which can make them reluctant to take independent actions without explicit approval. When this limitation is played by the opponent, the opponent picks one unit that is in command, and that unit functions as if it is out of command (dice for results as if the unit was outside of the leader's command radius).

Treacherous

The commander has a reputation for betrayal and cannot be fully trusted by their own troops and fellow leaders. When this limitation is played by the opponent, one unit attached to or adjacent to the leader must pass a Morale check. If the check is failed the unit functions as if it is out of command.

Unfamiliar Terrain

The commander is leading their forces in unfamiliar or hostile terrain, which hampers their ability to effectively coordinate their troops. When this limitation is played by the opponent, the affected leader's units suffer a -1 penalty to Movement during that turn, representing the difficulty of maneuvering through unfamiliar terrain.

Outnumbered

The commander is leading a force that is significantly outnumbered by the enemy, which can make it difficult to lead an army to achieve its objectives. When this limitation is played by the opponent, the affected leader's unit suffers a -1 penalty to all rolls during that phase. This Limitation can be applied to attack rolls, defense or morale rolls. The Outnumbered commander can use his Inspiration to counter this effect which represents the difficulty of facing overwhelming numbers on the battlefield.

Image courtesy of JC Lira.

Alexander's Upbringing

The story of Alexander the Great starts with his father, Philip II of Macedon. Philip II was a king who ruled the ancient kingdom of Macedonia from 359 to 336 BCE. Philip is remembered today for his military innovations, his skills as a diplomat, and the way he grew the Macedonian kingdom, which set the stage for Alexander the Great's later victories.

Philip II entrusted the education of his son Alexander to the renowned philosopher Aristotle. In 343 BCE, when Alexander was around 13 years old, Aristotle was invited to the Macedonian court at Pella to serve as a tutor for the young prince. Aristotle was already a well-known philosopher who studied under Plato, so he was an excellent choice for this important job.

Alexander's education under Aristotle was comprehensive and covered a wide range of subjects. He studied philosophy, ethics, politics, poetry, drama,

> **Formations, Arms & Armor of Alexander's Era**
>
> In subsequent chapters we'll detail the differences between the Greek phalanx, Macedonian phalanx and various support troops in Hellenistic armies. Additionally, we'll explain the tactics and arms of Alexander's adversaries.

and the natural sciences, such as biology, zoology, and astronomy. Aristotle also encouraged Alexander's love of Homer's works, especially the Iliad, which had a profound impact on how Alexander saw heroes and what he wanted to do with his own life.

Alexander also received a great deal of military training from a young age. He learned how to ride a horse,

Image courtesy of JC Lira.

shoot an arrow, and do other combat skills. This combination of intellectual pursuits and military discipline helped shape Alexander into a well-rounded and highly capable leader who would later achieve remarkable success in his military campaigns and conquests. Alexander probably learned how to fight, lead, and make peace by watching and taking part in his father's political and military campaigns. This would have laid the foundation for Alexander's own military and political successes later in his life.

Philip II was considered an excellent military tactician. He changed the way the Macedonian army worked by giving them new tactics, training, and weapons that were crucial to their success. The Macedonian phalanx, a tight formation of infantry with soldiers holding long sarissa pikes, was his most important innovation. This formation gave infantry more reach and protection, and it worked very well in battle. Philip was not only good at tactics, but also good at strat-

Alexander's Adversaries

In subsequent chapters we'll detail both the Macedonians and their Hellenistic allies as well as the nations and tribes that opposed Alexander's quest for glory.

egy. He knew when to use diplomacy, alliances, and marriages to get what he wanted.

Philip II was also involved in political intrigue, both within Macedonia and in the wider Greek world. He was able to deal with the complicated web of alliances and rivalries between Greek city-states, using their differences to grow Macedonia's power. Additionally, he was known to use diplomacy, bribery, and other non-military means to achieve his goals when possible. His marriage to Olympias, a member of the powerful Molossian royal house in Epirus, was a strategic move to secure an alliance with her influential family.

Philip II fought several adversaries during his reign, including neighboring Greek city-states like Athens, Thebes, and Thessaly, as well as the Illyrians and the Thracians. He was able to bring many of these states under his control or make them his allies, employing a powerful alloy of military power and diplomatic maneuvering. The Macedonian phalanx was Philip's most important military innovation. It was a tactical formation that made his infantry more effective in battle.

Philip's assassination in 336 BCE ended his reign. The assassination took place in 336 BCE during the wedding celebrations of Philip's daughter, Cleopatra of Macedon (not to be confused with the famous Egyptian queen), to her uncle, Alexander I of Epirus. The assassin, a man named Pausanias of Orestis, was a member of the elite Macedonian bodyguard corps.

Although the precise motives for his murder are still unknown, many historians think that Pausanias was acting out of personal grievances. Some historians think that the assassination may have been part of a larger plot involving members of the Macedonian court or even foreign powers. However, this is still a matter of historical debate.

Chaeronea (338 BCE)

The Battle of Chaeronea took place in 338 BCE near the city of Chaeronea in Boeotia, a region of

---

A word of caution about battles featured in this game:
Never learn history from a game.

Particularly from this game!

Alexander & His Adversaries includes a handful of conflicts that are fascinating for a variety of reasons in each section of the era of Alexander the Great. These notes can help you figure out how to build up that fight to play a game. Be aware that these setups are inspired by history rather and they are not intended as simulations of actual events.

Furthermore, because the topic is ancient history, unlike modern-day conflicts, evidence can be sparse. In some instances, what actually occurred during the battle is a subject of academic debate. Even academic views about some of these conflicts are based on scant evidence and some conjecture and experience.

More importantly, these scenarios have had total changes to reality from time to time and without notice to create what we believe to be a more enjoyable and/or fair game. Take them for what they are: a battle game and not a history lesson.

Alexander & His Adversaries

Image courtesy of Kevin Krause.

ancient Greece. This battle was a decisive engagement between the forces of Philip II of Macedon and a coalition of Greek city-states led by Athens and Thebes.

The exact numbers of troops involved in the battle are uncertain, and estimates vary among historians. But most historians think that Philip II's army, which included infantry and cavalry, had between 30,000 and 40,000 soldiers. The Macedonian phalanx made up a large part of the infantry. They were armed with long sarissa pikes and arranged in a tight formation that was both offensive and defensive. The elite Companion cavalry was also a key part of Philip's military strategy.

On the other side, the Greek city-state coalition was made up of contingents from Athens, Thebes, and several other city-states. The size of their combined forces was likely comparable to that of the Macedonian army, with estimates ranging from 30,000 to 40,000 soldiers. The Greek army was mostly made up of hoplites, who were heavily armed infantrymen with spears and big round shields who fought in the traditional phalanx formation. The Theban army also included the elite Sacred Band, a unit of 300 highly trained and dedicated soldiers.

Philip's plan involved feigning a retreat to draw the enemy forces out of position, then launching a devastating counterattack. In the middle of the battle, Alexander, Philip's 18-year-old son, led the elite Companion cavalry in a decisive charge that broke up the Theban Sacred Band, an elite infantry unit. This showed great courage and tactical skill. The victory at Chaeronea solidified Philip's control over Greece and demonstrated the potential of the young Alexander, who would go on to become one of the most successful military Leaders in history.

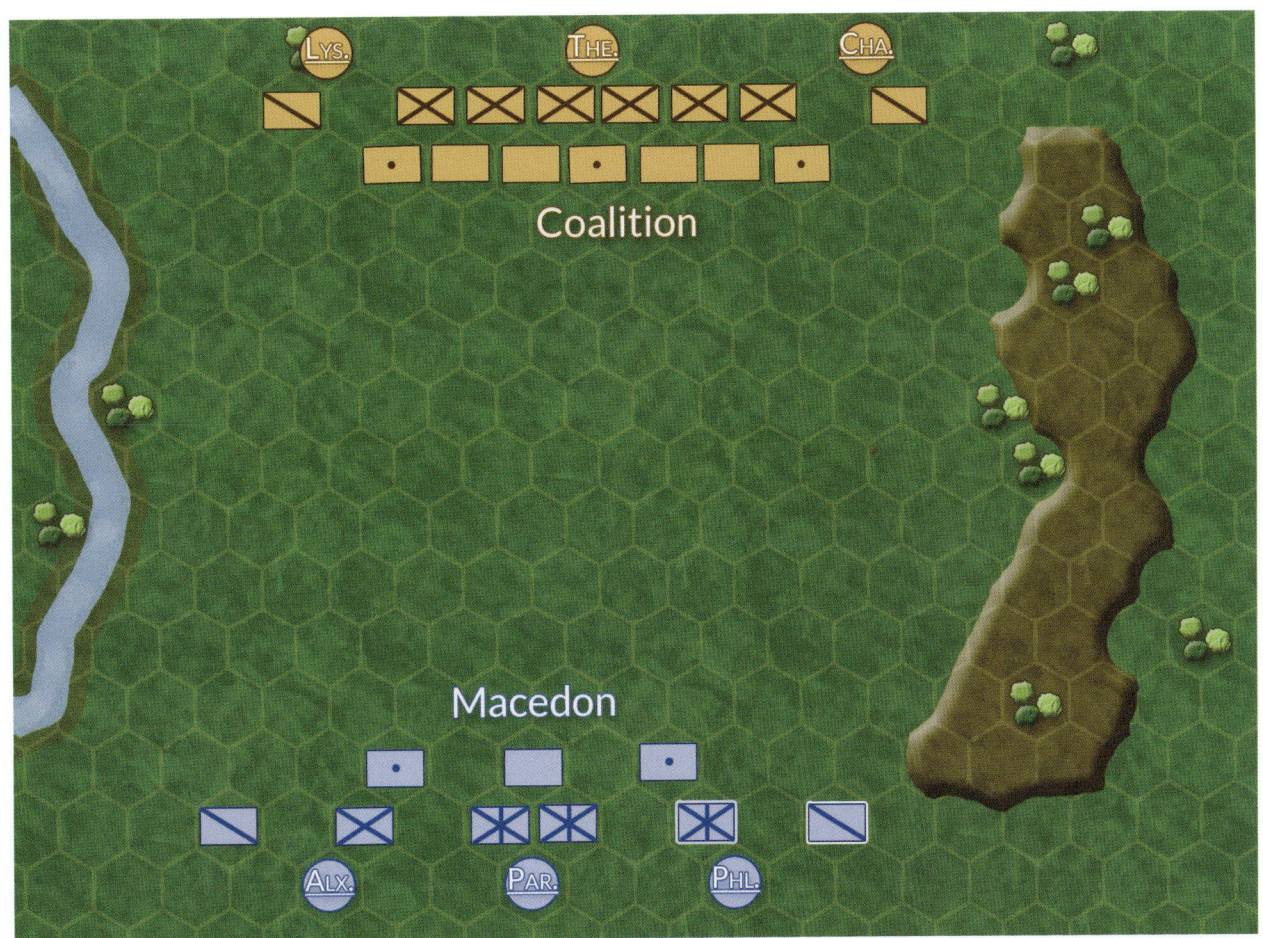

ALEXANDER & HIS ADVERSARIES

Macedon

Number	Type	Strength	Discipline	Vulner.	Whiff	Ranged
2	Psiloi	4	0	5	0	*slings*
1	Peltasts	4	0	4	1	*javelins*
1	Hoplites	8	1	3	1	
2	Phalangites	10	1	2	0	
1	Pezhetairoi	12	2	2	-1	
1	Light Cavalry	4	2	2	0	*javelins*
1	Companion Cavalry	4	4	2	-2	

Leader	Leadership	Tactical Advantage	Value	Inspirations	Limitations
Philip II	5	3	27	Innovator of the Phalanx	Fatherly Concern
Prince Alexander	4	3	24	Alexander's Charge	Brash
Parmenion	3	2	17	Tactical Flexibility	Aspiring Leader

Coalition

Number	Type	Strength	Discipline	Vulner.	Whiff	Ranged
3	Psiloi	4	0	5	0	*slings*
4	Peltasts	4	0	4	1	*javelins*
6	Hoplites	6	0	3	1	
2	Light Cavalry	4	2	2	0	*javelins*

Leader	Leadership	Tactical Advantage	Value	Inspirations	Limitations
Theagenes	4	1	17	Seize the Initiative, Rally	Aspiring Leader
Chares	2	1	10	Standard	Aspiring Leader
Lysicles	2	1	10	Standard	Aspiring Leader

Alexander & His Adversaries

Alexander's Armies

Alexander's infantry and cavalry worked together on the battlefield more than they had in the past. This approach is exemplified by Alexander the Great's use of the Macedonian phalanx alongside the Companion Cavalry, which allowed him to achieve remarkable success in his campaigns against the Persian Empire and beyond. The innovations and adaptations of Alexander would carry over to the Hellenistic armies of Successor Wars and beyond. Hellenistic use of cavalry and infantry during would have a lasting impact on the warfare of later civilizations.

Image courtesy of Kevin Krause.

Alexander & His Adversaries

Image courtesy of JC Lira.

Hoplites

During the era of Philip II, the hoplite was the primary infantryman in the armies of the various Greek city-states. Hoplites were citizen-soldiers, usually middle-class men who could afford to provide their own arms and armor.

Hoplite relied on the strength of fighting in a shield-to-shield formation to present a unified front and overpower the enemy through a combination of mass and discipline. Hoplites would advance towards the enemy at a steady pace, with the goal of engaging them in close combat. The phalanx worked best when the soldiers could stay in a tight formation on flat ground. However, it was vulnerable to attacks from the side and less effective on rough terrain or when the tight formation was broken.

Greek Phalanx vs Macedonian Phalanx

Hoplites fought in a group called a phalanx, which was usually made up of eight rows of soldiers close together. However, in these rules, to avoid confusion, the term phalanx is reserved for those phalangites with sarissas following the Macedonian tradition established by Philip and perfected by Alexander. So even though many troops fight shoulder-to-shoulder in a phalanx formation, only the Macedonian/sarissa phalanx is called a phalanx.

As citizen-soldiers, hoplites did not undergo the same level of professional training as the Macedonian phalanx. Still, they received some basic training and took part in regular drills to improve their skills and keep their bodies in shape. This training was mostly about learning how to stay in formation, move as a group, and use their weapons well in battle. Hoplites worked well together and were effective on the battlefield because they felt a sense of community, camaraderie, and shared civic duty.

While the Greek hoplite phalanx was a formidable force in and of itself, Philip II's innovations in the Macedonian phalanx eventually outclassed it and had a significant impact on the shifting power dynamics in the Greek world.

Weapons & Armor

Greek hoplites were equipped with several key weapons and armor. The Hoplon (or Aspis) is a large, round shield made of wood and covered with bronze, measuring around 30 to 40 inches (76 to 100 cm) in diameter. The shield was used to protect the hoplite and the soldier to his left.

Helmets

Ancient Greek and Mediterranean helmets can be categorized into several types based on their design, period, and region of origin. Here are some of the most common types:

Corinthian Helmet: This type of helmet was popular during the Archaic and early Classical periods (8th to 5th centuries BC). It's named after the city-state of Corinth and is one of the most iconic designs of ancient Greek armor. It is made of bronze and covers the entire head and neck, with slits for the eyes and mouth, and features a pronounced nose guard. The Corinthian helmet provides excellent protection but limited visibility and hearing. It was typically worn tipped back on the head when not in combat.

Chalcidian Helmet: This helmet style emerged in the 5th century BC and is named after the region of Chalcis on the island of Euboea. It's a modification of the Corinthian design, featuring larger eye holes and an open face for improved visibility and hearing. It often included cheek guards that could be pushed out of the way and a smaller nose guard. The Chalcidian helmet sometimes featured ornamental designs, such as a crest holder.

Phrygian or Thracian Helmet: This helmet style is named for the regions of Phrygia and Thrace. The design was influenced by the Phrygian cap, which has a forward-pointing peak. The Thracian helmet often had a large visor for increased protection, and it sometimes featured additional elements like cheek pieces and a top crest. The Thracian design was prevalent from the 5th to 1st centuries BC.

Pilos Helmet: This helmet is named after the pilos, a type of conical hat worn in ancient Greece. The pilos helmet was typically made of bronze and covered the top of the head, with the sides open for the ears. It offered less protection than other designs but was simpler to produce and more comfortable to wear. It was used from the 5th to 4th centuries BC, often by light infantry and archers.

Each type of helmet offered a different balance of protection, visibility, and comfort, reflecting the diverse approaches to warfare and equipment in the ancient world. Keep in mind that the names we use for these helmets are modern designations; the ancient Greeks didn't necessarily categorize their equipment in the same way.

The primary weapon of the hoplite was a spear (called a dory) measuring around 7 to 9 feet (2.1 to 2.7 meters) in length, significantly shorter than the Macedonian sarissa (below). The spear had an iron spearhead for thrusting and a bronze buttspike at the other end, which could be used as a secondary weapon or to anchor the spear in the ground. A secondary weapon, usually a short xiphos or kopis, used for close combat when the spear was no longer practical.

Hoplites wore various types of armor, including a bronze cuirass or linen linothorax to protect the torso, a bronze helmet to protect the head, and greaves to protect the lower legs.

Shield Wall

Most armies of antiquity can form a Shield Wall. Shield Walls are mostly defensive and they offer better protection against ranged weapons. In this reference, only medium and heavy infantry units can form a shield wall. Skirmishers may not form a shield wall.

- **Speed** - A Shield Wall has 2 Movement rather than 3 found in most other infantry
- **Flank** - A Shield Wall can "refuse the Flank". When a Unit refuses the flank, the formation curves to protect itself from flank attacks. A flank attack on a unit with a Shield Wall is treated as a front attack on a normal unit.

Image courtesy of New Buckenham Historical Wargamers.

Alexander & His Adversaries

Image courtesy of Richard Evers.

- **Limited Attack** - When dicing attacks, the Shield Wall rolls a maximum of 5 attack dice. The Shield Wall has no flanks, so it may attack to its sides, but to the side the Shield Wall can roll a maximum of 3 dice.
- **Protection From Ranged Weapons** - The Shield Wall provides a bonus of +1 to Vulnerability versus thrown weapons and projectiles.

The Sarissa Phalanx

Philip II of Macedon created the sarissa phalanx. Philip's phalanx was a highly effective infantry formation that became the foundation of the Macedonian army and has advantages and disadvantages over other infantry formations which we will describe below. Philip's phalanx was an adaptation of the traditional Greek phalanx, but with several key innovations in tactics, arms, and training that made it a formidable force on the battlefield.

The sarissa phalanx was characterized by its deep and tight formation, typically arranged in ranks of 16 men deep or more. This compact arrangement of phalangites provided better protection and cohesion, allowing the phalanx to act as a single, unified body. The phalanx was designed to maintain a steady advance, using its long sarissas to keep the enemy at bay and gradually push them back. However, the phalanx was less effective in rough or broken terrain, where its tight formation could be disrupted.

Philip II placed great emphasis on the training and discipline of his soldiers. The phalangites underwent rigorous drills to ensure they could maintain formation, maneuver, and fight effectively as a cohesive unit. Phalangites practiced moving in unison, maintaining the proper distance between ranks, and handling their sarissas with precision. The Macedonian phalanx was well-trained and disciplined to the point where it could do complicated moves and still be effective in battle.

Image courtesy of Richard Evers.

The phalanx was a key part of the military successes of Philip II and, later, Alexander the Great. Philip referred to his elite phalangites as *Pezhetairoi*, which translates as "foot companions," emphasizing the phalanx's significance to the monarch. Some of Alexander's elite sarissa troops were the *Hypasists*, or "shield bearers." After Alexander's death, the Hypaspists were reformed into a unit known as the "Silver Shields" (Argyraspides) by his Successors. The Argyraspides, many of them veterans of Alexander's campaigns, would play a significant role in the wars of the Diadochi that followed Alexander's death.

Philip's innovations in tactics, arms, and training allowed the Macedonian army to overcome numerous adversaries and secure the kingdom's dominance over the Greek world.

Weapons & Armor

One of the most significant innovations of the phalanx was the introduction of the sarissa, a long pike measuring between 18 and 22 feet (5.5 and 6.7 meters) in length. This was a substantial increase compared to the shorter spears used by the traditional Greek hoplites. The sarissa had a sharp iron spearhead at one end and a bronze buttspike at the other, which could be used to anchor the pike in the ground or as a secondary weapon if needed. The long reach of the sarissa allowed the first few ranks of the phalanx to present a deadly wall of spear points, making it difficult for the enemy to approach.

The secondary weapon of the Macedonian phalanx was a short sword, typically one of two types: the

Image courtesy of Richard Evers.

xiphos or the kopis. The xiphos was a double-edged, leaf-shaped short sword made of iron or bronze. It was primarily used for thrusting and stabbing in close combat when the long sarissa was no longer practical or had been broken. The kopis, on the other hand, was a single-edged, curved short sword with a forward-curving blade, designed for slashing and chopping. It was also effective in close combat when the sarissa could not be used.

The soldier's choice of secondary weapon could depend on his or her own preferences or the resources that are available. Both the xiphos and the kopis were good close-quarters weapons, and they were reliable backups when the main weapon, the sarissa, wasn't suitable or wasn't available.

Phalangites typically wore a simple helmet made of bronze or iron, designed to protect the head while allowing for good visibility and ventilation. Some examples include the Thracian, Phrygian, or Boeotian-style helmets, which were lighter and more open than the helmet used by Greek hoplites.

For body protection, phalangites often wore a linen cuirass called a linothorax, made from layers of glued or stitched linen. The linothorax was lighter and more flexible than a metal cuirass. It gave decent protec-

Image courtesy of New Buckenham Historical Wargamers.

tion without making it hard to move around. Some phalangites might have also used bronze or iron cuirasses, though these were less common due to their weight and cost.

Phalangites might wear bronze or leather greaves, similar to what Greek hoplites wore, to protect their lower legs. However, not all phalangites used greaves, as they could be cumbersome and impede movement.

Macedonian phalangites carried a smaller shield called a pelta or pelte, made of wood and covered with leather or bronze. The shield was slung across the shoulder with a strap, leaving both hands free to hold the sarissa. The smaller size of the shield provided less overall protection than the larger hoplon used by Greek hoplites, but it was lighter and allowed for greater mobility.

The armor and equipment of the Macedonian phalangites were designed to balance protection with the need for mobility and flexibility in the phalanx formation. Their gear allowed them to effectively wield their long sarissas and maintain the cohesion and effectiveness of the phalanx on the battlefield.

Advantages & Disadvantages

- **Speed** - A phalanx has 2 Movement rather than 3 found in most other infantry
- **Flank** - A phalanx cannot attack to its flank. A flank attack on a phalanx is treated as a rear attack on a normal unit.

Image courtesy of Richard Evers.

- **First Attack** - When dicing attacks, roll the phalanx's attacks first (except against an opposing phalanx). After the phalanx' attacks are resolved, roll the opponent's attacks. In other words, the great reach of the sarissa means that opponents must suffer Disruptions and Casualties before they can reply.
- **Rough Ground** - On rough terrain, a phalanx is at a distinct disadvantage against more flexible infantry. This weakness is abstracted as follows. Firstly, the phalanx never gets a first attack. Furthermore, the Phalanx must roll "Initiative" at the start of each Melee Phase. If the Phalanx wins or ties Initiative, then combat is considered simultaneous. If the Phalanx loses Initiative, then the opposing infantry has a First Attack on the Phalanx.
- **Forest** - A Phalanx cannot enter a forest.

The First Attack rule means that the phalanx is very nearly indestructible to the front. And historically, the phalanx was an indomitable force of nature until the Roman maniple exposed its weakness.

Peltasts

A peltast was a type of light infantryman in ancient Greek warfare, named after the small shield they carried, called a pelta or pelte. Peltasts were typically armed with a range of missile weapons, such as javelins or slings, and served as skirmishers and support troops in battle. They were highly mobile and used hit-and-run tactics, harassing the enemy from a distance before retreating to avoid direct engagement.

In general, peltasts wore lighter armor than hoplites, which made it easier for them to move quickly and keep their agility on the battlefield. Addition-

Alexander & His Adversaries

Image courtesy of Kevin Krause.

ally, the Persians used scythed chariots pulled by horses and manned by an archer or spearman and a driver. Their armor often consisted of a simple tunic or leather armor, with little to no head or leg protection. In addition to their missile weapons, peltasts might carry a short sword or dagger for close combat if necessary.

The role of peltasts was to weaken and disrupt enemy formations before the main infantry forces engaged in close combat. They would target vulnerable troops, such as archers or light infantry, and create confusion in the enemy ranks. In some cases, peltasts could also be used to pursue and harass retreating enemy forces.

Image courtesy of Richard Evers.

Image courtesy of JC Lira.

Psiloi

Psiloi (singular: psilos) were a type of light infantry in ancient Greek warfare, often serving as skirmishers and support troops. They were typically armed with missile weapons such as slings, bows, or javelins, and their primary role was to harass and weaken enemy forces from a distance before the main battle lines engaged in close combat.

Compared to other infantry types like hoplites or peltasts, psiloi wore little to no armor, which allowed them to move quickly and maintain their agility on the battlefield. They generally wore simple tunics or basic clothing and may have carried a small shield for protection.

The job of psiloi was to disrupt and harass enemy formations by going after archers, light infantry, and other psiloi. Their missile attacks could cause casualties, break enemy morale, and create confusion in the enemy ranks. Psiloi were also used to screen the main infantry forces, providing a buffer against enemy skirmishers and missile troops.

In some situations, psiloi could be used to chase and harass retreating enemy forces, taking advantage of their superior mobility to stop the enemy from regrouping or getting away. Even though they weren't as well armed or armored as other types of infantry, the psiloi were very important in ancient Greek warfare because of how they fought and how they helped other troops.

Image courtesy of New Buckenham Historical Wargamers.

Skirmish

Psiloi are skirmish troops. Skirmishers are low density troops designed to harass and blunt an enemy attack. Skirmishers typically are lightly armored and may carry projectiles or thrown weapons. Skirmishers never stand in shoulder-to-shoulder dense formations like infantry. Rather, skirmishers are disbursed to maximize reliance upon the individual's personal valor and discretion.

Skirmishers cannot hold ground against units other than Skirmishers. Skirmishers may not charge any unit except a Skirmisher. Skirmishers do not disrupt allies when they Fall Back or Retreat.

- **Speed** - A Skirmish unit has 4 points of Movement rather than 3.
- **Flank** - A Skirmisher has no flanks or rear. A Skirmish Unit allows individuals in the unit discretion to turn to face attacks on any side.
- **Limited Strength** - Skirmish units cannot have more than 4 Troops per unit.
- **Cannot Hold Ground** - If a Skirmish unit is charged by a unit other than a Skirmish unit, the Skirmish unit will always Fall Back.
- **Expected to Fall Back** - A Skirmisher does not cause disorder in allied units when it Falls Back or Retreats. A Skirmisher does not take any Disruptions when it falls back.
- **No Recovery** - Skirmishes cannot recover troops.

Image courtesy of JC Lira.

Cavalry

Hellenistic cavalry of the era, particularly during the reign of Alexander the Great and in the years following his death, underwent significant changes and improvements. The Macedonian Companion Cavalry (Hetairoi) was the best and most famous of these cavalry units. They were very important to Alexander's military campaigns.

Companion Cavalry

The Companion Cavalry was the elite heavy cavalry unit of the Macedonian army, personally led by Alexander the Great in battle. They were primarily composed of the Macedonian nobility and were armed with long lances (xyston), swords. They wore helmets (often Boeotian type) and cuirasses for protection. The primary role of the Companion Cavalry was to deliver devastating charges, often targeting enemy Leaders or breaking through the enemy lines. The combined use of Companion Cavalry with phalanx was often called the anvil and hammer. The phalanx is the anvil to pin the enemy, and the companions swing round to hammer the blow. The Companion Cavalry's discipline, skill, and ability to maneuver on the battlefield made them a formidable force.

Image courtesy of Kevin Krause.

Thessalian Cavalry

During the Hellenistic period, the Thessalian cavalry was another important cavalry force. It came from the area of Thessaly in northern Greece. They were renowned for their skill in horsemanship and were typically deployed as medium or heavy cavalry. Armed with spears or lances, they were used for both shock combat and skirmishing.

Image courtesy of Richard Evers.

Image courtesy of Richard Evers.

Light Cavalry

The era also saw the use of light cavalry, which included various types of skirmishers and scouts. These units were typically armed with javelins, or xyston and played a significant role in reconnaissance, harassing the enemy, and protecting the flanks of the main army. Soldiers from different ethnic groups, like the Thracians, Illyrians, and Paeonians, often made up the majority of the light cavalry.

Spearhead Formation

The Mounted Spearhead, Spearhead Formation, or more commonly the *Spearhead*, is a tactical formation employed by some of the most elite cavalry units of the time, such as the Companion Cavalry and Thessalian Cavalry. Only well-disciplined mounted troops capable of maintaining the tight formation and adapting to the dynamic nature of the battlefield can form Spearhead.

The Spearhead is a mass of tightly packed cavalry massing in a pointed formation resembling a spearhead. The purpose of this formation is to maximize the impact and penetration capabilities of the cavalry, allowing them to break through enemy lines or exploit weaknesses in the opposing force's formation. While the Spearhead offers several advantages in terms of offensive power, it also has its vulnerabilities and limitations.

Forming Spearhead or resuming normal order costs 1 Movement point. This action can be taken during the Movement Phase of a player's turn, when the cavalry unit is activated. The formation change is considered part of the unit's movement for the turn and should be accounted for when planning the unit's movement and actions. It's essential for players to consider the tactical implications of changing formations and to carefully manage their units' Movement points to optimize their effectiveness on the battlefield.

Advantages & Disadvantages

Charge Bonus: When charging into melee combat, a Spearhead formation gains a +2 bonus to its Attack value.

Armor Penetration: The focused force of a Spearhead formation allows it to better penetrate enemy armor. When in this formation, the unit gains a +1 bonus to its Attack value against heavily armored opponents.

Breakthrough Potential: If a Spearhead formation successfully eliminates an enemy unit in melee combat, it may immediately make a bonus move of up to 2 hexes in a straight line.

Alexander & His Adversaries

Formation Vulnerability: A Spearhead formation is more vulnerable to flank and rear attacks. The unit suffers an additional -1 penalty to its Defense value when attacked from the flank or rear.

Limited Maneuverability: The tight formation of a Spearhead limits its ability to maneuver. When in this formation, the unit's Movement is reduced by 1.

Discipline Requirement: A unit must have a Discipline of at least 3 to be able to form a Mounted Spearhead formation.

Greco-Macedonian Army List

The Greco-Macedonian Army List represents the various military forces during time before Philip's death and during Alexander the Great's reign. This army list is diverse and flexible, reflecting the mix of native and foreign troops utilized by the Macedonian army. It can be used as a generic Macedonian list under Alexander himself, showcasing the unique blend of units and tactics that made his army so successful on the battlefield.

Greco-Macedonian						
Type	Strength	Discipline	Vulner.	Whiff	Ranged	Value
Hoplites	8	1	3	1		17
Phalangites	10	1	2	0		24
Pezhetairoi	12	2	2	-1		31
Peltasts	4	0	4	1	*javelins*	11
Psiloi	4	0	5	0	*slings*	19
Light Cavalry	4	2	2	0	*javelins*	31
Thessalian Cavalry	4	3	2	-1		27
Companion Cavalry	4	4	2	-2		30
Heavy Cavalry	4	2	1	-1		27

Macedonia's primary leader during this time is King Philip II.

Leader	Leadership	Tactical Advantage	Value	Country	Inspirations	Limitations
Philip II	5	3	27	Macedon	Innovator of the Phalanx	Fatherly Concern
Prince Alexander	4	3	24	Macedon	Alexander's Charge	Brash
Parmenion	3	2	17	Macedon	Tactical Flexibility	Aspiring Leader
Antipater	3	2	17	Macedon	Reliable Commander	Aspiring Leader
Cleitus the White	2	1	10	Macedon	Standard	Aspiring Leader
Ptolemy	2	1	10	Macedon	Standard	Aspiring Leader
Perdiccas	2	1	10	Macedon	Standard	Aspiring Leader
Craterus	2	1	10	Macedon	Standard	Aspiring Leader

The coalition forces are ruled by consensus and oppose Philip and/or Alexander.

Leader	Leadership	Tactical Advantage	Value	Country	Inspirations	Limitations
Theagenes	4	1	16	Thebes	Tactical Flexibility	Aspiring Leader
Chares	2	1	10	Athens	Standard	Aspiring Leader
Lysicles	2	1	10	Athens	Standard	Aspiring Leader
Memnon of Rhodes	3	1	13	Rhodes	Seaborne	Aspiring Leader
Stratocles	2	1	10	Athens	Standard	Aspiring Leader
Skofeldes *	3	2	17	Illyria	Guerrilla	Aspiring Leader
Bardylis	2	1	10	Illyria	Guerrilla	Aspiring Leader

Fictionalized leader

ALEXANDER & HIS ADVERSARIES

Image courtesy of Kevin Krause.

Alexander Becomes King

Alexander the Great became king of Macedonia following the assassination of his father, Philip II. After the assassination, Alexander, then 20 years old, was quickly declared king by the Macedonian army and the nobles. However, he faced immediate challenges to his rule, including rival claimants to the throne and uprisings in various regions of the kingdom. Alexander acted decisively to consolidate his power and eliminate potential rivals. He had several family members and other potential claimants executed to secure his position and quell any rebellions.

Alexander also needed to establish his legitimacy and authority in the eyes of his subjects and the wider Greek world. As a young king, he had to prove himself as a capable ruler and military Leader, capable of upholding his father's legacy and expanding the kingdom. His early successes, both in dealing with internal threats and rebellions and in leading impressive military campaigns, helped to solidify his reputation as a strong and effective leader.

By 335 BCE, Alexander had control over his neighbors, so he focused on maintaining their loyalty and fostering unity among them. He encouraged the various Greek city-states to join the League of Corinth, a federation of Greek states that recognized Alexander as their hegemon and military leader. In return, Alexander guaranteed the autonomy of the city-states in their internal affairs. This alliance was important for Alexander's later plans, as it provided a united front and a pool of resources for his upcoming campaigns.

The Cynic

The meeting of Alexander the Great and Diogenes the Cynic, recounted by historians including Diogenes Laërtius and Plutarch, compares the stark contrast between the pursuit of worldly power and the philosophy of the Cynic. Cynicism, a school of thought in ancient Greece, taught that the path to virtue and happiness was to live in simplicity and naturalness, rejecting social conventions and material possessions.

In the story, Alexander visits Diogenes in Corinth. Diogenes, living in accordance with Cynic philosophy, had abandoned all worldly goods and was residing in a pithos (a large ceramic jar). When Alexander, showing respect for the philosopher, asked if there was any favor he could grant him, Diogenes simply replied, "Yes, stand out of my sunlight."

Alexander was reportedly so impressed by Diogenes' independence and disregard for societal norms that he stated, "If I were not Alexander, then I should wish to be Diogenes."

This anecdote underscores the juxtaposition value of simplicity, self-sufficiency, and independence against the pursuit of wealth and power. Ultimately, so

Image courtesy of JC Lira.

Macedonian Army List

\multicolumn{7}{c}{Hellenic}						
Type	Strength	Discipline	Vulner.	Whiff	Ranged	Value
Hoplites	8	1	3	1		17
Phalangites	10	1	2	0		24
Foot Companions	12	2	2	-1		31
Peltasts	4	0	4	1	*javelins*	11
Psiloi	4	0	5	0	*slings*	19
Light Cavalry	4	2	2	0	*javelins*	31
Thessalian Cavalry	4	3	2	-1		27
Companion Cavalry	4	4	2	-2		30

Leader	Leadership	Tactical Advantage	Value	Country	Inspirations	Limitations
King Alexander	4	3	24	Macedon	Alexander's Charge	None
Craterus	2	1	10	Macedon	Disciplinarian	Subordinate Commander
Conus	1	1	7	Macedon	Master of Maneuver	Subordinate Commander
Meleager	1	1	7	Macedon	Unbreakable Will	Subordinate Commander
Perdiccas	2	2	14	Macedon	Steadfast Defender	Subordinate Commander
Ptolemy	2	1	10	Macedon	Adaptable Tactician	Subordinate Commander
Peithon	1	1	7	Macedon	Skilled Reconnaissance	Subordinate Commander

Thrace & Illyria

After Philip II was assassinated, tribes in the north like the Illyrians, the Thracians, and others rose up against Alexander. The young king acted quickly and marched his army north to subdue these rebellions. In 335 BCE, he launched a series of campaigns against the Thracians and Illyrians, defeating them in several battles and reestablishing control over these regions.

The exact locations of all the battles against the Thracians and Illyrians are not well documented, but a few key engagements are known. Alexander first dealt with the Thracian rebellion. One of the notable battles took place near the Haemus Mountains (in present-day Bulgaria). Alexander beat the Thracians by using both foot soldiers and mounted soldiers to counter their guerrilla-style tactics. This campaign demonstrated Alexander's tactical adaptability in dealing with unconventional warfare.

After subduing the Thracian tribes, Alexander turned his attention to the Illyrians. These events taught him important lessons that would help him win battles against the Persian Empire and other enemies in the future. These events taught him important lessons that would help him win battles against the Persian Empire and other enemies in the future.

Image courtesy of Kevin Krause.

Image courtesy of Kevin Krause.

Haemus Mountains (335 BCE)

While it is clear that Alexander campaigned against the Thracians and Triballians in the region around the Haemus Mountains, the specific details of the battles, participants, and numbers involved remain uncertain due to the lack of reliable and consistent historical sources.

According to some accounts, in 335 BCE, Alexander launched a campaign against the Thracians and Triballians in response to their rebellion following Philip II's death. The Thracians and Triballians were led by various chieftains and tribal leaders.

Under Alexander, the Macedonian army was usually made up of the phalanx, which was armed with long sarissas; the elite Companion cavalry; different types of light infantry like archers, peltasts, and psiloi; and other auxiliary forces, depending on the situation and the resources available. The Macedonian forces were well-trained and had good discipline. They also used advanced strategies and formations that helped them win.

On the other hand, the Thracian and Triballian forces were likely composed of various tribes and chieftains, each with their own warriors and fighting styles. They usually used smaller infantry units, used guerrilla-style tactics, and took advantage of the

Image courtesy of Kevin Krause.

fact that the locals knew the area well. These forces would have included skirmishers armed with javelins, slings, or bows, and melee infantry with spears or swords. Because there aren't many historical records, it's hard to know exactly what forces were at this battle.

Cersobleptes and Skostokos are placed a little differently from most games. Skostokos places his units anywhere adjacent to the village. Cersobleptes units are held in reserve and appear as per the ambush ability of the Guerilla Inspiration. That is to say, they are placed on the board in or adjacent to any forest hex during any Movement Phase. That placement is the unit's movement. The Thracian player can place any 4 units from the reserve under either commander.

The Macedonian player wins if at the end of a turn, Macedon occupy 1 of the 3 village hexes, while the Thracian player does not occupy any village hex. The Thracian player wins if he can drive 2 phalanxes out of the game or deny the village to the Macedonian for 12 turns.

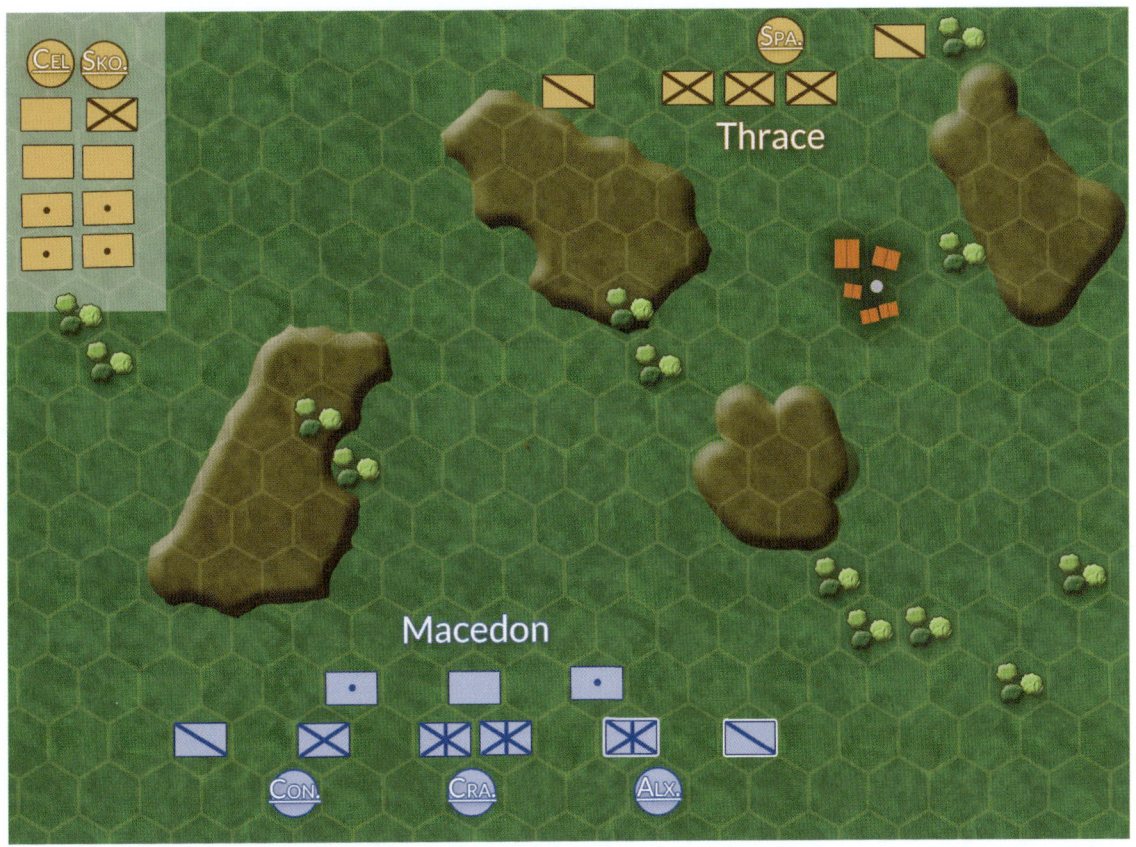

Image courtesy of Richard Evers.

Macedon

Number	Type	Strength	Discipline	Vulner.	Whiff	Ranged
2	Psiloi	4	0	5	0	slings
1	Peltasts	4	0	4	1	javelins
1	Hoplites	8	1	3	1	
2	Phalangites	10	1	2	0	
1	Foot Companions	12	2	2	-1	
1	Light Cavalry	4	2	2	0	javelins
1	Companion Cavalry	4	4	2	-2	

Leader	Leadership	Tactical Advantage	Inspirations	Limitations
King Alexander	4	3	Alexander's Charge	None
Craterus	2	1	Disciplinarian	Aspiring Leader
Conus	1	1	Master of Maneuver	Aspiring Leader

Thrace & Illyria

Number	Type	Strength	Discipline	Vulner.	Whiff	Ranged
4	Hoplites	6	0	3	1	
3	Peltasts	4	0	4	1	javelins
4	Psiloi	4	0	5	0	slings
2	Light Cavalry	4	2	2	0	javelins

Leader	Leadership	Tactical Advantage	Inspirations	Limitations
Seuthes II	2	1	Mountain Fighter	Aspiring Leader
Cersobleptes	1	1	Guerrilla	Aspiring Leader
Birtelos *	2	1	Skirmish Expert	Aspiring Leader

Fictionalized leader.

Image courtesy of Richard Evers.

Pelium

At Pelium, King Skofeldes of the Illyrians and Glaukias of the Taulantii of the Dardanians led a coalition against young King Alexander. We don't know exactly how many soldiers were in the Battle of Pelium because there aren't many written accounts of it and the ones that do exist don't agree with each other. However, Alexander is said to have commanded a

force of around 15,000–20,000 men, while the Illyrian and Dardanian coalition likely had a comparable or slightly smaller number of troops.

The Illyrian and Dardanian armies were probably made up of many different clans and chieftains, each with their own warriors and ways of fighting. They usually used smaller infantry units, used guerrilla-style tactics, and took advantage of the fact that the locals knew the area well. These forces would have included skirmishers armed with javelins, slings, or bows, and melee infantry with spears or swords. The exact composition of their forces is difficult to determine due to the limited historical accounts available.

MACEDON

Number	Type	Strength	Discipline	Vulner.	Whiff	Ranged
1	Psiloi	4	0	5	0	slings
1	Peltasts	4	0	4	1	javelins
1	Hoplites	8	1	3	1	
1	Phalangites	10	1	2	0	
1	Pezhetairoi	12	2	2	-1	
1	Light Cavalry	4	2	2	0	javelins
1	Companion Cavalry	4	4	2	-2	

Leader	Leadership	Tactical Advantage	Inspirations	Limitations
King Alexander	4	3	Alexander's Charge	None
Langarus	4	2	Loyal Friend	Loyalties

ILLYRIA

Number	Type	Strength	Discipline	Vulner.	Whiff	Ranged
4	Hoplites	6	0	3	1	
5	Peltasts	4	0	4	1	javelins
4	Psiloi	4	0	5	0	slings

Leader	Leadership	Tactical Advantage	Inspirations	Limitations
Skofeldes *	3	2	Guerrilla	Aspiring Leader
Glaukias of the Taulantii	3	1	Skilled Reconnaissance	Local Loyalties

* Fictionalized leader.

Alexander & His Adversaries

Rivals & Usurpers

The rivals and usurpers army list represents Thracians, Illyrians, and other factions that challenged the kingship of Alexander. This list is characterized by a mix of traditional Hoplites, skilled Guerillas, versatile Peltasts, and mobile Psiloi and Light Cavalry. Guerillas have the capabilities of Peltast infantry but they are also capable of moving through Rough Terrain as if it were Open Terrain.

This army excels at exploiting difficult terrain and utilizing hit-and-run tactics to outmaneuver their opponents. The combination of various ranged capabilities and diverse troop types enables this army to adapt to a wide range of battlefield situations and exploit the weaknesses of a more traditional Hellenic army such as Philip or Alexander might field.

Rivals and Usurpers						
Type	Strength	Discipline	Vulner.	Whiff	Ranged	Value
Hoplites	6	0	3	1		12
Peltasts	4	0	4	1	*javelins*	11
Psiloi	4	0	5	0	*slings*	19
Light Cavalry	4	2	2	0	*javelins*	31

Leader	Leadership	Tactical Advantage	Value	Country	Inspirations	Limitations
Menmon of Rhodes	3	1	14	Persia	Inspiring Presence	Aspiring Leader
Masizej *	2	1	10	Persia	Guerrilla	Aspiring Leader
Cersobleptes	1	1	7	Thrace	Guerrilla	Aspiring Leader
Seuthes II	2	1	10	Thrace	Mountain Fighter	Aspiring Leader
Stefasebis *	3	2	20	Illyria	Steadfast	Cautious
Birtelos *	2	1	10	Thrace	Skirmish Expert	Aspiring Leader
Glaukias of the Taulantii	3	1	13	Taulantii (Illyrian)	Skilled Reconnaissance	Local Loyalties

* Denotes a fictionalized name.

Alexander & His Adversaries

Image courtesy of Kevin Krause.

Alexander Invades Persia

Alexander turned his attention to the Persian Empire once he had control over his Hellenistic neighbors. In 334 BCE, he started a campaign to conquer Persia and fulfill a long-held Greek wish to avenge Persian invasions of Greece in the past. Philip II was already preparing for this invasion. By executing Philip's plan, Alexander sought to turn his own kingdom into an empire by overthrowing Darius III, sovereign ruler of the vast Persian Empire.

In 334 BC, Alexander the Great began his invasion of Persia with 30,000 infantry and 5,000 cavalry entering Asia Minor at the Hellespont (now known as the Dardanelles). The Hellespont was the gateway to Asia Minor (now known as Turkey).

The first significant battle of the campaign unfolded near the Granicus River in northwestern Asia Minor. A Persian army under the command of local satraps (governors) and Greek mercenaries met Alexander's

Image courtesy of Richard Evers.

The Gordian Knot

While advancing through Asia Minor, Alexander arrived in the city of Gordium, where an ancient prophecy stated that whoever could untie the complex knot binding a cart to a post would become the ruler of Asia.

Many had tried and failed to untie the knot, but Alexander was determined to fulfill the prophecy. According to one account, after examining the knot and realizing its complexity, Alexander drew his sword and cut through it, thus "solving" the riddle and securing his destiny as the ruler of Asia. This story highlights Alexander's determination, ambition, and ingenuity, which would become defining features of his rule and conquests.

forces. In a decisive victory, Alexander managed to defeat the Persian forces and pave the way for further conquests in Asia Minor.

Following the Battle of Granicus, Alexander continued his campaign, moving southward along the Aegean coast. His next major engagement was the Siege of Halicarnassus in 334 BC. Here, the Persian leadership appointed Memnon of Rhodes, a skilled Greek mercenary, to defend the city. Even though Alexander was able to take Halicarnassus in the end, Memnon and his troops put up a fierce fight that destroyed the city but showed how

Image courtesy of Richard Evers.

Image courtesy of Kevin Krause.

determined the Persian leaders were to stop the invasion.

When King Darius III heard about Alexander's victories, he put together a massive army to defeat Alexander's Macedonians. This culminated in the Battle of Issus in 333 BC, which took place in southeastern Asia Minor near the modern-day border between Turkey and Syria. The Persian forces were outnumbered, but Alexander's tactics and leadership led to a crushing defeat for the Persians.

As Alexander's forces emerged victorious in the battle, Darius III had no choice but to flee, leaving his mother, wife, and children behind. Upon discovering the Persian royal family, Alexander chose to treat them with great respect and kindness, rather than as prisoners of war. He ensured that they were provided with proper care and comfort, even allowing them to retain their royal status.

When Darius III learned of Alexander's noble treatment of his family, he was said to have been moved and grateful. According to some ancient sources, Darius even offered Alexander a sizable ransom for their release, along with a peace agreement that included sharing power in the Persian Empire. Alexander, however, declined Darius's offer, as he had no intention of sharing power. Alexander was set on conquering the entire Persian Empire.

GRANICUS (334 BC)

The Battle of Granicus in 334 BC marked the beginning of Alexander's Persian campaign. The conflict unfolded near the Granicus River in northwestern Asia Minor. The Macedonian forces comprised heavy infantry arranged in a phalanx formation, the elite Companion Cavalry, light infantry, including archers and slingers, and light cavalry. The Persian forces consisted of a mix of Persian infantry, Greek mercenaries in hoplite formation, light infantry, archers, and a strong cavalry component. The terrain near the river was relatively flat, with some slopes near the riverbank. Alexander's tactics involved crossing the river and launching a surprise attack on the Persian cavalry, then targeting the Greek mercenaries with his heavy infantry. The Macedonian phalanx and the swift assault of the Companion Cavalry led to a decisive victory for Alexander.

Regarding the map, not that the sources tell different tales. Arrian, Plutarch and Diodorus Siculus wrote about the battle, with Diodorus having a different account telling of the crossing of the whole army before the battle. But other two have more detailed accounts, with heroic fighting across the river. Writers like John Warry and Nik Sekunda prefer the fight across the river.

If you prefer, run the stream through the middle of the field between the armies.

Alexander & His Adversaries

Macedon

Number	Type	Strength	Discipline	Vulner.	Whiff	Ranged
4	Phalangites	10	1	2	0	
2	Pezhetairoi	12	2	2	-1	
2	Hypastists	6	3	4	1	
2	Psiloi	4	0	5	0	*slings*
2	Thessalian Cavalry	4	3	2	-1	
1	Companion Cavalry	4	4	2	-2	
1	Light Cavalry	4	2	2	0	*javelins*

Leader	Leadership	Tactical Advantage	Inspirations	Limitations
Alexander the Great	6	3	Tactician, Veteran Commander	None
Parmenion	5	2	Veteran Commander	Age

Persians

Number	Type	Strength	Discipline	Vulner.	Whiff	Ranged
2	Kardakes (Hoplite mercenaries)	8	0	3	1	
2	Takabara (Peltast mercenaries)	4	0	4	1	*javelins*
4	Psiloi	4	0	5	0	*slings*
8	Light Cavalry	4	2	2	0	*javelins*

Leader	Leadership	Tactical Advantage	Inspirations	Limitations
Arsites	3	0	Standard	Cowardice
Menmon of Rhodes	3	1	Inspiring Presence	Aspiring Leader
Mithradates	2	1	Standard	
Spithridates	1	1	Standard	
Rhoesaces	1	1	Standard	

Alexander & His Adversaries

Issus (333 BC)

The Battle of Issus in 333 BC took place in southeastern Asia Minor, near the modern-day border between Turkey and Syria. Both armies featured a diverse array of troops. The Macedonians had their heavy infantry phalanx, Companion Cavalry, light infantry, and light cavalry, while the Persians had a large force of infantry, archers, heavy and light cavalry, chariots, and Greek mercenaries.

Darius and Alexander clashed on a narrow coastal plain that bordered the sea to the west and mountains to the east. This limited space neutralized the Persians' numerical advantage. Alexander's strategy hinged on a well-timed cavalry charge led by himself, which targeted the Persian left flank and forced King Darius III to flee the battlefield. Meanwhile, the Macedonian phalanx held the line against the Persian infantry, resulting in a significant victory for Alexander.

Image courtesy of JC Lira.

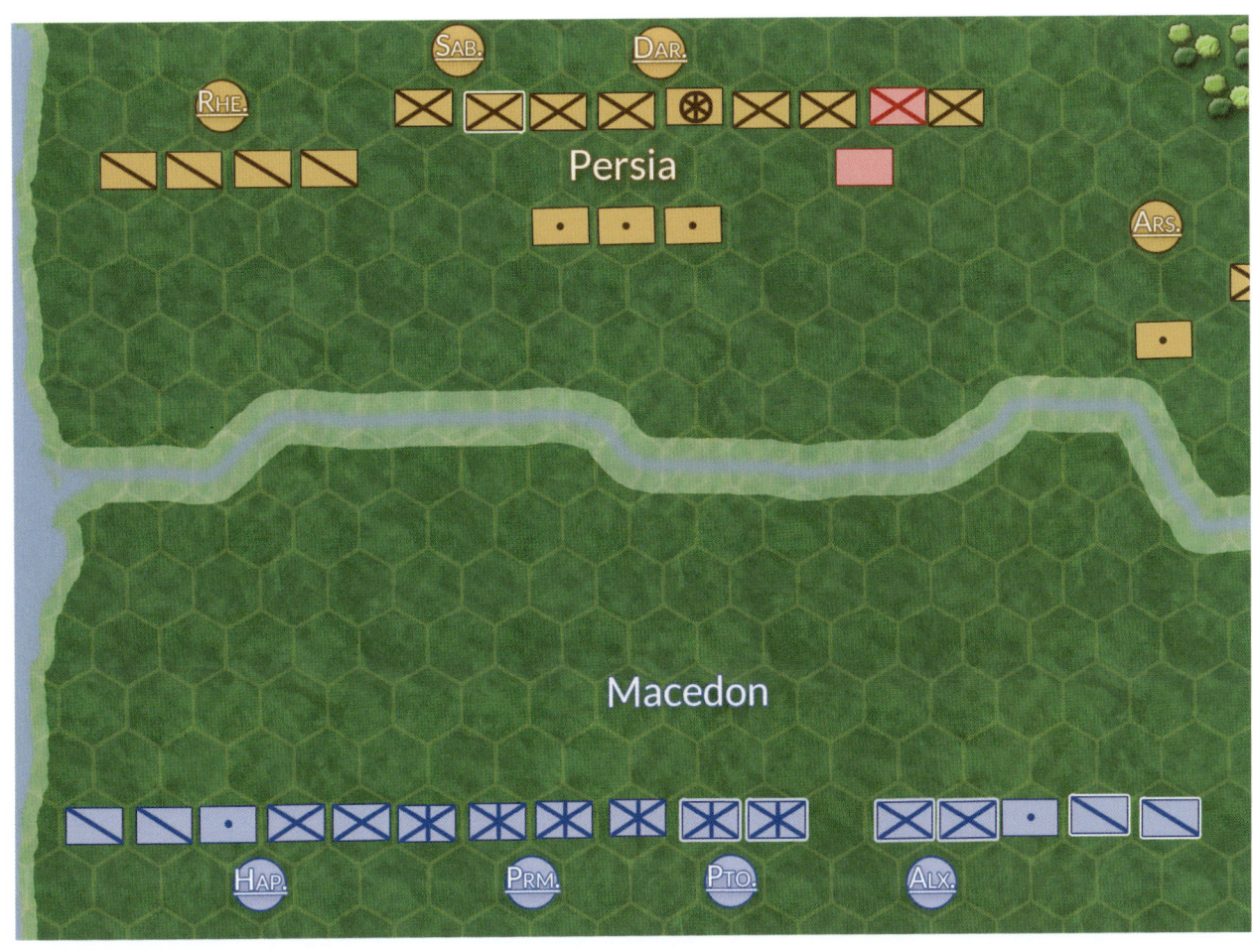

The brook at the center of the battlefield is considered rough terrain and is easily fordable at all points.

Alexander & His Adversaries

\	MACEDON					
NUMBER	TYPE	STRENGTH	DISCIPLINE	VULNER.	WHIFF	RANGED
4	Phalangites	10	1	2	0	
2	Pezhetairoi	12	2	2	-1	
2	Hypastists	6	3	4	1	
2	Psiloi	4	0	5	0	slings
2	Thessalian Cavalry	4	3	2	-1	
1	Companion Cavalry	4	4	2	-2	
1	Light Cavalry	4	2	2	0	javelins

LEADER	LEADERSHIP	TACTICAL ADVANTAGE	INSPIRATIONS	LIMITATIONS
Alexander the Great	6	3	Tactician, Veteran Commander	None
Parmenion	5	2	Veteran Commander	Age
Hephaestion	4	2	Loyal Friend	Loyalties
Ptolemy	4	1	Tactician	Aspiring Leader

\	PERSIANS					
NUMBER	TYPE	STRENGTH	DISCIPLINE	VULNER.	WHIFF	RANGED
1	Hoplite Mercenary	8	0	3	1	
1	Peltast Mercenary	4	0	4	1	javelins
4	Skirmish	4	0	5	0	bows or slings
4	Light Cavalry	4	2	2	0	javelins
1	Scythed Chariot	4	2	0	1	
7	Sparabara	6	1	4	0	javelins
2	Immortals	8	2	3	-1	bows

LEADER	LEADERSHIP	TACTICAL ADVANTAGE	INSPIRATIONS	LIMITATIONS
Darius III	4	1	Royal Authority	Fear of Defeat
Arsames	4	1	Resourceful	Loyalties
Sabaces	4	1	Steadfast	Fear of Defeat
Rheomithres	4	1	Standard	none

Army of Alexander the Great

Type	Strength	Discipline	Vulner.	Whiff	Ranged	Value
Hoplites	8	1	3	1		17
Phalangites	10	1	2	0		24
Pezhetairoi	12	2	2	-1		31
Peltasts	4	0	4	1	*javelins*	11
Psiloi	4	0	5	0	*slings*	19
Light Cavalry	4	2	2	0	*javelins*	31
Thessalian Cavalry	4	3	2	-1		27
Companion Cavalry	4	4	2	-2		30
Heavy Cavalry	4	2	1	-1		27

Leader	Leadership	Tactical Advantage	Value	Country	Inspirations	Limitations
Alexander the Great	6	3	45	Macedonia	Tactician, Veteran Commander	None
Parmenion	5	2	23	Macedonia	Veteran Commander	Age
Hephaestion	4	2	20	Macedonia	Loyal Friend	Loyalties
Ptolemy	4	1	24	Macedonia	Tactician	Aspiring Leader
Craterus	4	1	16	Macedonia	Reliable Commander	Cautious
Antigonus	4	1	16	Macedonia	Tactician	Aspiring Leader
Perdiccas	4	1	16	Macedonia	Steadfast	Ambitious
Seleucus	4	1	16	Macedonia	Resourceful	Aspiring Leader

Alexander & His Adversaries

Image courtesy of JC Lira.

Achaemenid Persia

During the era of Alexander the Great, the Achaemenid army, led by King Darius III, employed a diverse array of troops and tactics. The size of the Achaemenid Persian Empire was reflected in its army, which was made up of soldiers from many different ethnic groups and military traditions. The Achaemenid army relied on the fact that it had more soldiers, that its cavalry was flexible, and that its archers were good.

The Persians aimed to use their cavalry to outflank and harass the enemy, while their infantry and archers provided a solid defensive line. However, in battles against Alexander's highly disciplined and innovative Macedonian army, these tactics were often not sufficient to secure victory. The combination of Alexander's superior tactics, the effectiveness of the Macedonian phalanx, and the elite Companion cavalry often proved to be a decisive advantage against the Persian forces.

Sparabara

The Persian infantry, known as sparabara, were the backbone of the Achaemenid army. The term "sparabara" derives from Old Persian and means "shield-bearers" because these warriors carried large rectangular wicker shields into battle.

As Persians and Medians were the dominant ethnic groups within the Achaemenid Empire, they comprised the majority of the sparabara. However, it should be noted that the Persian Empire was vast and diverse, and as a consequence, soldiers from other conquered regions may have been integrated into

---

Achaemenid Empire

The term "Achaemenid" refers to the Achaemenid Empire, also known as the First Persian Empire, which was founded in the sixth century BCE by Cyrus the Great. The name Achaemenid is taken from Achaemenes (Haxmani in Old Persian), the ruling dynasty's eponymous ancestor. Achaemenes, also known as Hakhamanish, was most likely a Persian tribe chieftain or local ruler in the early 7th century BCE.

The Achaemenid Empire was a vast and powerful empire that included territories in modern-day Iran, Iraq, Turkey, Egypt, Afghanistan, and parts of Central Asia, the Caucasus, and the Indus Valley. The Achaemenid Empire was the world's largest empire at its peak, both in terms of geographical extent and people.

From Cyrus the Great's founding in 550 BCE until Darius III's defeat in 330 BCE, the Achaemenid dynasty ruled the kingdom. The empire was renowned for its effective administration, ethnic diversity, tolerance, and infrastructure initiatives such as the Royal Road, which facilitated communication and trade throughout the vast empire. The Achaemenid Persian Empire set the groundwork for the region's subsequent empires and had a significant impact on the growth of art, architecture, and governance throughout the ancient world.

sparabara units. There could be soldiers from Babylon, Lydia, or even Egypt among them.

As well-trained and disciplined combatants, the sparabara had a generally high level of morale and effectiveness. They constituted the front line of the Persian army and were responsible for defending archers and other light infantry units. The sparabara were armed with spears, swords, and large shields, allowing them to engage in close combat or form a shield wall against hostile missile attacks.

During the Achaemenid Empire, the sparabara were effective in many battles, but they struggled against opponents with heavier armor or greater discipline, such as the Greek hoplites or the Macedonian phalanx led by Alexander the Great.

Immortals

The Immortals were an elite element of the military forces of the ancient Persian Achaemenid Empire. They were referred to as "Immortals" because their

Image courtesy of Kevin Krause.

Alexander & His Adversaries

number was always maintained at 10,000; if a soldier was slain or injured, he was promptly replaced, creating the illusion of an immortal and unstoppable force. The Immortals were predominantly Persian and Median soldiers. Originally, they functioned as the bodyguard of the Persian king but by the time of Alexander they had evolved into an elite infantry force.

Archers

The Achaemenid Persian army was renowned for its skilled archers, who played a significant role in battles. They used composite bows, which had a longer range and greater accuracy compared to simple wooden bows. The archers often formed a line behind the infantry, providing supporting fire.

Image courtesy of JC Lira.

Alexander & His Adversaries

Cavalry

The Persian cavalry was another essential component of the army. They were divided into light and heavy cavalry units. Light cavalry was typically armed with javelins and bows, while heavy cavalry, often Persian nobles, wore armor and wielded lances or spears. The Persian cavalry, a predecessor of cataphracts, used protective coverings for their horses, often made of leather or metal plates. These coverings were designed to safeguard the horse's vital areas, such as the chest, neck, and flanks. They were typically adorned with decorative elements to enhance the visual appearance. The Persian cavalry excelled in hit-and-run tactics and flanking maneuvers.

Scythed Chariots

The Persians utilized scythed chariots, which were manned by an archer or spearman and a driver and pulled by horses. These chariots were equipped with scythe blades mounted on the wheels or axles, intended to cause significant damage to enemy infantry or cavalry. However, by the time of Alexander the Great, their effectiveness in battle had diminished, and they were primarily used for psychological impact. References to scythed chariots in battle are found in the armies of Persians, Seleucids, and Pontic forces. Although the Persians employed scythed chariots against Alexander, his well-disciplined and adaptable army ultimately overcame the Persian forces. While

Image courtesy of JC Lira.

Alexander & His Adversaries

> ### *Chariots at Guagamela*
>
> During the Battle of Gaugamela, the Persians deployed scythed chariots as part of their strategy. The chariots were intended to disrupt the tightly organized Macedonian phalanxes and create chaos in their ranks.
>
> However, Alexander had studied the Persian tactics and devised countermeasures. He anticipated the threat posed by the scythed chariots and came up with strategies to neutralize their impact. Prior to the battle, Alexander ordered his men to widen the gaps between their shield walls to allow the chariots to pass harmlessly through their ranks. He also instructed his soldiers to target the charioteers themselves rather than the horses or the chariots.
>
> When the battle commenced, the Persian scythed chariots charged towards the Macedonian lines, but Alexander's troops held their ground and allowed the chariots to pass through the gaps. The Macedonians then targeted the exposed charioteers, killing or capturing them. This tactic proved successful, as the scythed chariots failed to achieve their intended purpose of disrupting the Macedonian phalanxes.

initially fearsome, the importance of scythed chariots dwindled over time as military tactics and technology progressed.

Elephantry

Even though they weren't used very often during Alexander's time, the Achaemenid Persian army did have war elephants, which were usually used to intimidate and disrupt enemy formations.

War elephants played a significant role in ancient military strategy, notably within the campaigns of Alexander the Great and his successors. The primary species utilized was the Asian Elephant (*Elephas maximus*), employed by both Persian and Indian armies.

Under Philip II of Macedonia, the Greco-Macedonian military strategy relied heavily on infantry formations, notably the phalanx. War elephants were largely unfamiliar to their forces until the reign of Alexander the Great.

During his eastern campaigns, Alexander encountered Persian armies using war elephants as mobile platforms for archers and shock troops. At the battle of Gaugamela for example, there were 15 elephants present. This approach disrupted enemy formations and often had a psychological impact, causing fear and disorder.

As Alexander's conquests led him into India, he met armies that had further mastered the use of elephants in warfare. The Battle of Hydaspes was a notable encounter where Alexander faced King Porus's force, which included a significant contingent of war elephants. Despite initial challenges, Alexander's forces developed tactics to counter this threat, focusing on the elephants' limited maneuverability.

Following Alexander's death, his successors, the Diadochi, maintained and expanded the use of war elephants. The Seleucid Empire, which controlled territories in the east, possessed a significant elephant corps. Despite their impact on the battlefield, the use of war elephants also presented logistical challenges, including control, management, and transportation.

Types of Elephant

In the era of Alexander the Great and his successors, the use of elephants in warfare was an established military strategy. The species employed ranged from the Asian Elephant, or the Indian elephant, to the now-extinct North African elephant, with their usage largely dictated by geographical availability and practicality in training.

The most commonly used was the Asian Elephant due to its wide distribution across India and Southeast Asia, regions where Alexander expanded his campaigns and where his successors maintained their territories. Asian elephants, standing about 9 to 10 feet tall, were revered for their strength and endurance. They served as formidable platforms for archers and shock troops, capable of trampling enemy lines and causing significant disarray.

A subspecies of the Asian elephant called the Syrian elephant, or Western Asiatic elephant was found in the region from Anatolia (modern day Turkey), throughout Syria, to as far east as the lands of the Indus River. Syran were smaller than the African elephants, but larger than the Asian elephants from India. Some scholars debate whether Syrian elephants were widely used in the ancient world for warfare, particularly in the armies of the Persians and later, the armies of Alexander the Great and his successors. It is possible that they were extinct before the time of Alexander.

The Seleucid Empire, one of the successor states after Alexander's death, was known to make extensive use of Syrian elephants in its armies. Seleucus I Nicator, the founder of the Seleucid Empire, is said to have received a number of these elephants in a peace treaty with the Maurya Empire in India, and used them to great effect in his campaigns.

It is believed that the Syrian elephants went extinct due to a combination of factors, including over-hunting, habitat loss, and capture for use in war. By the time of the Roman Empire, they had disappeared entirely from the region.

Another elephant used in antiquity was the North African elephant, also known as the Atlas elephant. The Atlas elephant was used less frequently but still played a prominent role in historical conflicts. Smaller than Asian elephants, they stood about 8 feet tall at the shoulder. They were noted for their maneuverability across various terrains. The Atlas elephant went extinct, likely due to a combination of overhunting—particularly for warfare and entertainment in Roman arenas—habitat loss from expanding human agriculture and settlements, and possible changes in climate and vegetation after the last ice age.

The usage of larger African elephants is virtually unheard of warfare during this era. Although significantly larger than both Asian and North African elephants, they are more aggressive and hence more challenging to train for battle purposes. The substantial investment of time and resources required to capture and train African elephants, coupled with their unpredictable nature, likely dissuaded their widespread use. Furthermore, their natural habitat was primarily outside the regions where elephant warfare was prevalent, limiting their accessibility to the armies of this period.

Mercenaries

The Achaemenid army often hired Greek mercenaries because they were very good at fighting with heavy infantry. These mercenary hoplites fought in the traditional Greek phalanx formation.

Weapons & Armor

During the era of Alexander the Great, the Achaemenid Persian Empire had a diverse range of weapons and armor, owing to its vastness and the various cultures it encompassed. Different types of Persian

Image courtesy of Kevin Krause.

troops had different weapons and armor that fit their roles in battle.

The primary weapon for the Persian infantry, particularly the sparabara, was the long spear, which they used for thrusting and maintaining a defensive line. Persian infantrymen, including the elite Immortals, carried short swords or daggers known as akinakes as a secondary weapon used for close combat.

Persian archers were renowned for their skill, and they used composite bows made from layers of wood, horn, and sinew. These bows had a longer range and greater accuracy than simple wooden bows. Archers also carried a quiver with a variety of arrowheads, like ones that could go through armor or ones with a wide tip to cause more damage.

Light cavalry often wielded javelins, which they threw at the enemy while maintaining mobility on the battlefield. Some light cavalry units also used composite bows, similar to those of the infantry archers, for hit-and-run tactics. But heavy cavalry, including Persian nobles, carried lances or spears, designed to deliver devastating charges and penetrate enemy lines. As a secondary weapon, cavalry units carried short swords or daggers for close combat.

Persian soldiers wore various types of helmets made from metal, leather, or padded fabric. Helmets of-

Image courtesy of Richard Evers.

ten featured cheek guards and, in some cases, neck protection.

Persian infantry and heavy cavalry wore cuirasses made of scale armor, leather, or padded fabric. Scale armor, made of small, overlapping metal scales, was flexible and provided good protection. Some elite units, like the Immortals, wore metal breastplates for additional protection.

The Persian infantry, especially the sparabara, carried large, rectangular shields made of wicker or wood and covered with leather or fabric. These shields were great for protecting them from arrows and other projectiles. Cavalry units used smaller round or crescent-shaped shields made from wood, leather, or metal.

Some Persian soldiers wore greaves (leg protection) and vambraces (arm protection) made of metal, leather, or padded fabric, especially in elite units or heavy cavalry.

Persian Army List

Persians						
Type	Strength	Discipline	Vulner.	Whiff	Ranged	Value
Immortals	8	2	3	-1	bows	48
Sparabara	6	1	4	0	javelins	25
Skirmish	4	0	5	0	bows or slings	19
Lt. Cavalry	4	2	2	0	bows	23
Hvy. Cavalry	6	2	1	-1		18
Scythed Chariot	4	2	0	1		19
Kardakes (Hoplite mercenaries)	8	0	3	1		17
Takabara (Peltast mercenaries)	4	0	4	1	javelins	11
Elephantry	4	3	2	-1	bows	38

Leader	Leadership	Tactical Advantage	Value	Country	Inspirations	Limitations
Darius III	4	1	10	Persia	Royal Authority	Fear of Defeat
Bessus	4	1	6	Bactria	Tactician	Treacherous
Menmon of Rhodes	3	1	14	Persia	Inspiring Presence	Aspiring Leader
Mazaeus	4	2	20	Babylonia	Veteran Commander	Aspiring Leader
Orontobates	3	1	13	Persia	Steadfast	Aspiring Leader
Satibarzanes	4	1	16	Persia	Fierce	Aspiring Leader
Arsames	4	1	16	Persia	Resourceful	Loyalties
Oxathres	3	1	13	Persian/Media	Tactician	Aspiring Leader

Image courtesy of Richard Evers.

Founding of Alexandria

Following his victories in Asia Minor, Alexander decided to dissolve his navy and engage in a land conflict with Persia. Because of this, he shifted south to deny the Persian fleet access to ports in the Mediterranean. This brought Alexander's army to the threshold of Persia's African satrapy (province), Egypt.

The Egyptian people resented Persian rule, so they welcomed Alexander as a liberator. The Persian satrap of Egypt surrendered without putting up significant resistance, making the conquest of Egypt relatively peaceful compared to other campaigns.

While in Egypt, Alexander founded the city of Alexandria. Alexander envisioned this city as a key commercial and cultural hub, strategically located on the Mediterranean coast. He personally selected the site for the city, which was situated between the Mediterranean Sea and Lake Mareotis, near the Nile Delta. Alexandria was designed by the Greek architect Dinocrates, who laid out the city's streets in a grid pattern, a typical feature of Greek urban planning.

While in Egypt, Alexander also visited the Oracle of Ammon at the Siwa Oasis, where, according to some accounts, the oracle declared him to be the son of Zeus-Ammon, reinforcing his divine status and further legitimizing his rule.

How many Alexandrias?

Alexandria in Egypt would eventually become one of the most important cities in the Hellenistic world, serving as a center of learning, culture, and trade. The famous Library of Alexandria and the Lighthouse of Alexandria, one of the Seven Wonders of the Ancient World, were both built in this city, further solidifying its prominence.

Apart from Alexandria in Egypt, Alexander the Great is said to have founded or renamed around 20 cities in his own name during his extensive conquests. While some of these cities grew to be significant centers, others were more modest or short-lived settlements.

Some notable examples include Alexandria Eschate ("Alexandria the Farthest") in modern-day Tajikistan, Alexandria Bucephalous (named after Alexander's horse Bucephalus) in modern-day Pakistan, and Alexandria Arachosia (modern-day Kandahar) in Afghanistan. It is important to note that the number of cities founded or renamed by Alexander may vary depending on the sources and interpretations by historians.

Image courtesy of New Buckenham Historical Wargamers.

Eastern Persia

After Alexander took over Egypt in 332 BC, he returned to Asia to continue his invasion of the Persian Empire. Before Alexander entered Egypt, he laid siege to the heavily fortified island city of Tyre, which was part of the Persian Empire. Alexander overcame the city's defenses by building a causeway between the mainland and the island. The siege lasted seven months. The fall of Tyre marked an important victory and secured Alexander's rear as he advanced into the Persian heartland.

The Battle of Gaugamela in 331 BC effectively sealed the fate of the Persian Empire. The battle occurred near present-day Mosul, Iraq, in a flat, open plain chosen by Darius III to maximize the effectiveness of his large army, chariots, and cavalry. Darius suffered another crushing defeat at Gaugamela.

After the Battle of Gaugamela, Alexander marched unopposed into the great city of Babylon and then captured the wealthy city of Susa. Both cities surrendered without resistance, and their vast treasuries greatly enriched Alexander's coffers.

Alexander continued his campaign by capturing Persepolis, the ceremonial capital of the Persian Empire. After a brief resistance, the city fell, and Alex-

Image courtesy of Kevin Krause.

Image courtesy of Kevin Krause.

ander allowed his troops to loot the city. In an infamous act, Alexander ordered the burning of the royal palace, possibly as a symbolic gesture of revenge for the Persian invasion of Greece and the burning of the Athenian Acropolis in 480 BC.

The Persian Empire was all but defeated, but in 330 BC Alexander pursued Darius III into the easternmost satrapies of the Persian Empire. However, before Alexander could capture Darius, Darius was betrayed and killed by one of his own satraps, a man named Bessus. After usurping Darius, Bessus declared himself king. But Bessus' reign over the greatly diminished empire was short-lived. Alexander eventually captured and executed Bessus, thereby eliminating the last significant resistance from the Persian Empire.

Gaugamela (331 BC)

Alexander's forces included his heavy infantry phalanx, Companion Cavalry, light infantry, and light cavalry, while the Persians fielded a massive army with infantry, archers, heavy and light cavalry, chariots, war elephants, and Greek mercenaries. In terms of tactics, Alexander used a modified oblique order, re-

Image courtesy of Richard Evers.

fusing his left flank and concentrating his forces on the right, where he led a powerful charge. This tactic created gaps in the Persian line and allowed Alexander to break through, eventually targeting Darius himself. When Darius ran away from the battlefield, the Persian forces broke apart, and Alexander won. This made it possible for Alexander's conquest of the Persian heartland.

There are a few unique rules in this scenario. Firstly, Persian cavalry can move off the board. Once off the table, the cavalry must remain out of play for one turn. On the following turn, the cavalry can appear on any vacant full-sized hex at any edge of the battlefield. This represents the very real danger that Alexander faced of being completely encircled.

In addition, if the scythed chariots charge, Macedonian foot troops can forgo their counter-attacks and allow the chariots to pass through the hex. When this happens, the chariots are placed behind the defending unit and neither side takes casualties. This represents the training that Alexander's men had against chariots. Note that chariot cannot turn to face an enemy when attacked from the rear.

ALEXANDER & HIS ADVERSARIES

Macedon

Number	Type	Strength	Discipline	Vulner.	Whiff	Ranged
4	Phalangites	10	1	2	0	
2	Pezhetairoi	12	2	2	-1	
2	Hypastists	6	3	4	1	
2	Psiloi	4	0	5	0	*slings*
2	Thessalian Cavalry	4	3	2	-1	
2	Companion Cavalry	4	4	2	-2	
0	Light Cavalry	4	2	2	0	*javelins*

Leader	Leadership	Tactical Advantage	Inspirations	Limitations
Alexander the Great	6	6	Tactician, Veteran Commander	None
Parmenion	5	2	Veteran Commander	Age
Hephaestion	4	2	Loyal Friend	Loyalties
Ptolemy	4	1	Tactician	Aspiring Leader

Persians

Number	Type	Strength	Discipline	Vulner.	Whiff	Ranged
1	Hoplite Mercenary / Karadaka	8	0	3	1	
1	Elephantry	4	3	2	-1	*bows*
2	Skirmish	4	0	5	0	*bows or slings*
8	Light Cavalry	4	2	2	0	*javelins*
3	Scythed Chariot	4	2	0	1	
5	Sparabara	6	1	4	0	*javelins*
2	Immortals	8	2	3	-1	*bows*

Leader	Leadership	Tactical Advantage	Inspirations	Limitations
Darius III	4	1	Royal Authority	Fear of Defeat
Bessus	4	1	Tactician	Treacherous
Mazaeus	4	2	Veteran Commander	Aspiring Leader
Orontes II	4	1	Standard	none

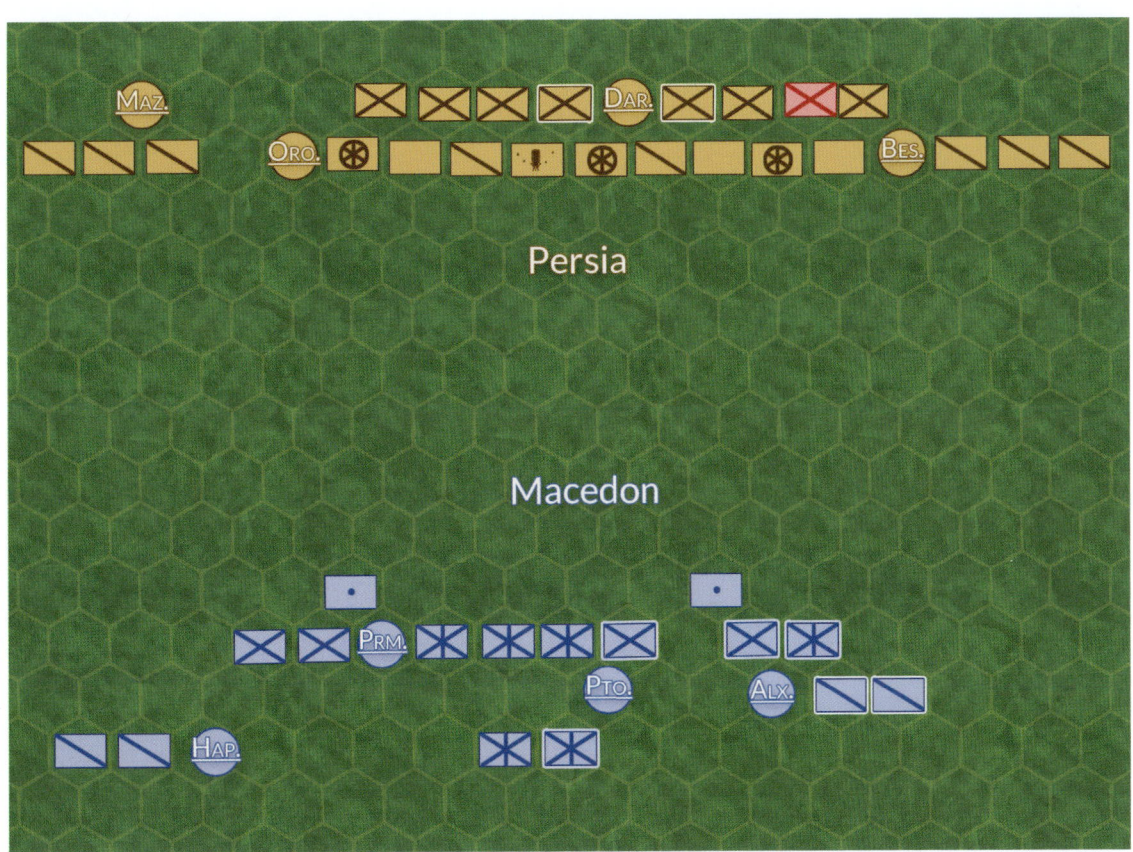

Image courtesy of Richard Evers.

ALEXANDER & HIS ADVERSARIES

What's the difference between heavy cavalry and cataphracts?

Bactria and Sogdiana did have heavy cavalry, and some of their forces may have been early forms of cataphracts. Cataphracts, as we know them today, were more closely associated with the Parthian and Sassanian Empires, which emerged later in the region's history.

Cataphracts were a type of heavily armored cavalry, often with scale or lamellar armor covering both the rider and the horse. They were armed with long lances called *kontos*, which they used to launch powerful shock charges into enemy lines. The word "cataphract" comes from the Greek word *kataphraktos*, which means "completely enclosed," and refers to the extensive armor protection.

Heavy cavalry, on the other hand, is a broader term for cavalry units that are better armored and equipped than light cavalry. Heavy cavalry may wear a variety of armor, ranging from chainmail to partial plate armor, and may be armed with lances, spears, or swords. Cataphracts are a type of heavy cavalry, but not all heavy cavalry are cataphracts.

The heavy cavalry Alexander faced during his invasion of Bactria and Sogdiana would not have been as heavily armored or specialized as the cataphracts that later emerged in the Parthian and Sassanian Empires. Bactrian and Sogdian heavy cavalry probably wore cataphract-style armor, such as scale or lamellar armor for the rider, but their horses were probably not as heavily armored as later cataphracts. They probably wore some kind of barding, like chaffron and peytrals.

Bactria and Sogdiana

After conquering the Achaemenid Persian Empire, Alexander turned his attention to the easternmost provinces of the empire, which included the regions known today as Afghanistan and parts of Central Asia.

In 330 BCE, Alexander's forces entered the territory of Bactria and Sogdiana, which are part of modern-day Afghanistan, Uzbekistan, and Tajikistan. The local people, collectively known as the Sogdians and Bactrians, were a mix of Iranian-speaking tribes and settled city-dwellers. Their armies were mainly composed of infantry, archers, and cavalry.

Bactrian and Sogdian infantry forces included both light and heavy infantry units. Typically, they were armed with spears, swords, and bows, and protected by wooden or wicker shields. For light infantry, their armor could range from simple tunics and leather to more advanced scale or lamellar. They fought in close formations similar to the Greek hoplites or Persian infantry. Archers, who were a significant part of these armies, used composite bows and were known for their precision and range.

Bactria and Sogdiana were renowned for their horse breeding, which resulted in a robust tradition

Roxana of Bactria

Spitamenes was a local Bactrian leader who led a tenacious and successful resistance to Alexander's troops during the Macedonian king's Central Asian campaign. Spitamenes used guerrilla warfare tactics and his knowledge of the local terrain to inflict significant damage on Alexander's troops while evading capture.

Spitamenes led a decisive victory over the Macedonians at the Battle of the Polytimetus River in 328 BCE, ambushing and defeating a Macedonian force headed by Alexander's general, Pharnuches. This defeat led Alexander to seize control of the situation and relentlessly pursue Spitamenes.

Alexander's campaign against Spitamenes lasted months, with the Macedonian monarch adapting to the Bactrian leader's guerrilla tactics. Under increasing pressure and with his forces dwindling, Spitamenes was eventually betrayed by his own men, who killed him and gave his head to Alexander in the winter of 328-327 BCE in the hope of ensuring their own safety.

During his chase of Spitamenes, Alexander met the noble Bactrian family of Oxyartes, whose daughter Roxana piqued the king's interest. Alexander married Roxana in a lavish ceremony that combined both Macedonian and Bactrian traditions in order to ensure the loyalty of the local Bactrian and Sogdian populations. This marriage was significant as it was the first time Alexander married a woman who was not Macedonian or Greek, and it marked a turning point in his policy of assimilating the people and culture of the countries he conquered. Roxana ultimately gave birth to him a son, Alexander IV, who became his sole legitimate heir.

of cavalry. They fielded both light and heavy cavalry units. Light cavalry was typically equipped with bows for hit-and-run tactics, whereas heavy cavalry was outfitted with long lances or spears and heavier armor for close combat. The heavy cavalry wore scale or chainmail armor and were armed with long lances.

In addition, during this campaign Alexander faced numerous nomadic and semi-nomadic tribes, including the Scythians and Massagetae, who excelled in horse archery and were renowned for their swift and highly mobile warfare.

Despite the fierce resistance, Alexander eventually subdued these regions and incorporated them into his growing empire. However, the difficulties he faced in Afghanistan and Central Asia foreshadowed the challenges his successors would encounter in maintaining control over these distant territories.

Bactria						
Type	Strength	Discipline	Vulner.	Whiff	Ranged	Value
Sparabara	8	0	4	0	*javelins*	32
Skirmish	4	0	5	0	*bows or slings*	19
Light Cavalry	4	2	2	0	*bows*	23
Heavy Cavalry	6	2	1	-1		32
Elephantry	4	4	2	-1	*bows*	38

Leader	Leadership	Tactical Advantage	Value	Country	Inspirations	Limitations
Spitamenes	3	2	17	Sogdiana	Guerrilla	Local Loyalties
Oxyartes	2	1	10	Bactria	Steadfast	Aspiring Leader
Bessus (Bactrian)	2	1	10	Bactria	Tactician	Unfamiliar Terrain
Datafernes	2	1	10	Sogdiana	Tactician	Aspiring Leader

To the Ends of the World

After conquering the Persian Empire, Alexander continued his eastward expansion, aiming to reach the "ends of the world and the Great Outer Sea." During this period, he sought to consolidate his rule, integrate the various conquered peoples, and explore new territories.

Integration and blending of cultures were a priority for Alexander after he had won the war. Alexander adopted some Persian customs and attire, encouraged intermarriage between his soldiers and local populations, and appointed Persians to administrative positions within his empire. He also held a mass wedding ceremony in Susa in 324 BC, where he and many of his officers married Persian noblewomen, symbolizing the union of Macedonian and Persian cultures.

Furthermore, Alexander pursued ambitious campaigns in the east. He invaded present-day Afghanistan, parts of Central Asia, and India. Alexander encountered fierce resistance and fought notable battles, such as the Siege of the Sogdian Rock in 327 BC, where he used skilled climbers to scale the seemingly impregnable fortress, surprising the defenders and securing a swift victory.

Image courtesy of Richard Evers.

Alexander and Kalanos

Kalanos, also known as Calanus or Kalyana, was an Indian philosopher and ascetic who lived during the time of Alexander the Great. He was a practitioner of the ancient Indian ascetic tradition known as "Gymnosophists," which translates to "naked philosophers." Calanus was a follower of the Sramana tradition, which included Jainism, Buddhism, and Ajivika philosophies, although it is not clear which specific school of thought he adhered to.

Alexander encountered Calanus during his invasion of India in 326 BC. Impressed by the wisdom and simplicity of the Indian sages, Alexander invited Calanus to join his court. Calanus agreed and became an adviser to Alexander, accompanying him during his campaigns in India and his return journey to Persia.

Calanus is said to have shared his philosophical insights with Alexander and his generals, influencing their understanding of Eastern thought. One famous anecdote tells of a conversation between Calanus and Alexander, in which the philosopher advised the conqueror to be a "ruler of men" rather than a "desirer of possessions," emphasizing the importance of virtue and wisdom over material wealth.

As they reached Persia, Calanus fell ill and decided to end his life through the traditional Indian practice of self-immolation, known as *sallekhana* or *samadhi-marana*. According to ancient accounts, he prepared a funeral pyre, offered his final teachings, and entered the fire willingly, all while maintaining a calm and meditative composure. Alexander and his court were deeply moved by this event.

It is said that the memory of Calanus remained with Alexander until the end of his life.

Alexander's invasion of India was marked by the famous Battle of the Hydaspes in 326 BC against King Porus, a powerful Indian ruler. Despite facing war elephants, Alexander's forces emerged victorious, and he subsequently treated Porus with respect, even allowing him to retain his kingdom as a vassal.

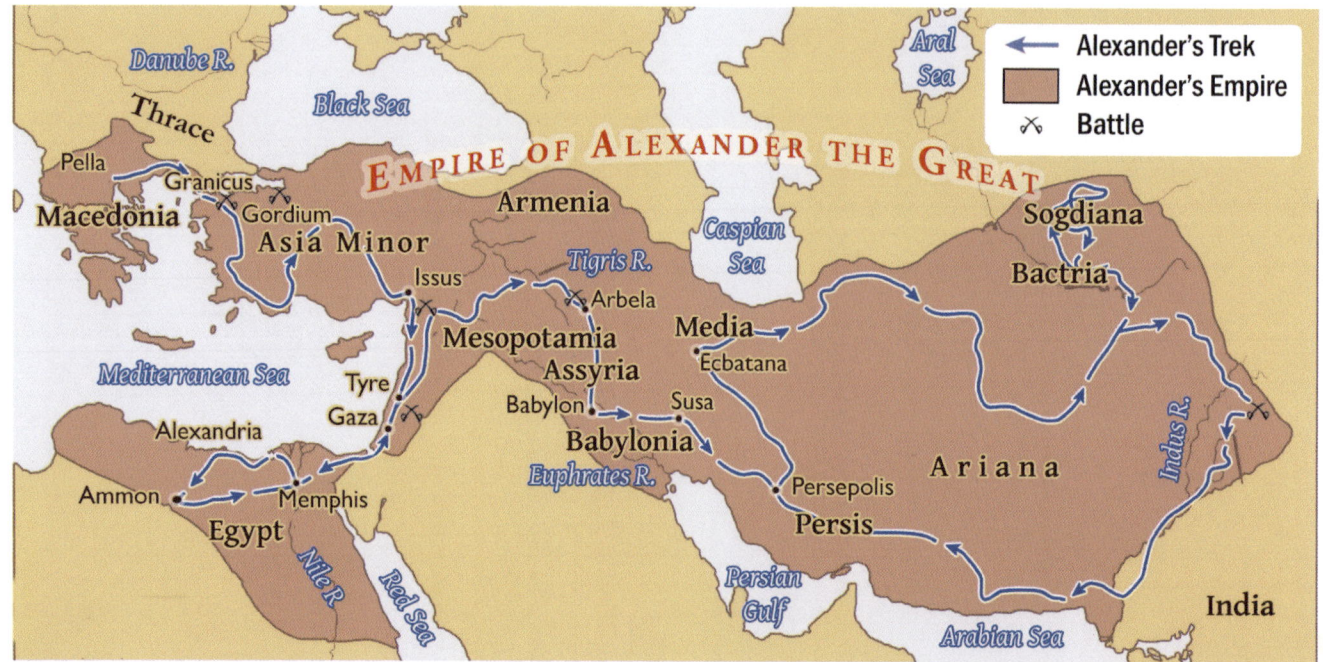

Alexander & His Adversaries

As Alexander continued his eastern campaigns, his troops eventually grew weary, and their morale waned. In 326 BC, when they reached the Hyphasis River (modern-day Beas River) in India, his soldiers refused to march further east. Alexander was forced to turn back and begin the long journey to Babylon, where he planned to consolidate his empire and launch new campaigns.

Did Alexander ever encounter the Chinese?

There is no direct evidence that Alexander the Great met or had a personal encounter with any Chinese people during his lifetime. However, because trade and cultural exchange occurred between China and the areas west of it, including the Persian Empire and various Central Asian states, Alexander may have had some indirect knowledge of China.

While Alexander did not directly meet or know any Chinese people, his conquests set the groundwork for the development of the Silk Road trade network, which would later become a major conduit for cultural exchange between the East and the West.

During Alexander's eastern campaigns, he conquered territories that stretched as far as modern-day Pakistan and Afghanistan for the Greeks at the time. He did not go any further east to investigate or conquer the lands leading to China. Nonetheless, his campaigns indirectly facilitated greater contact and trade between the Hellenistic world and the Orient.

Following Alexander's death, his conquests resulted in the establishment of the Hellenistic Seleucid Empire, which ruled over a vast region extending from Asia Minor to Central Asia. This kingdom interacted with the Greco-Bactrian Kingdom, a Hellenistic state in Central Asia, which in turn interacted with the Chinese via the Silk Road trade network. Furthermore, the Mauryan Empire, which arose in India after Alexander's death, maintained commerce and diplomatic ties with the Hellenistic states.

All of these links aided in the exchange of goods, ideas, and information between the Mediterranean world, Central Asia, India, and China. It is interesting to wonder what lands Alexander would have entered in Europe, Asia or Africa if he had not died so young.

Image courtesy of Kevin Krause.

India

When Alexander the Great invaded India in 326 BC, he encountered a different kind of military force than he had faced in Persia and Asia Minor. The Indian armies, particularly those of the powerful Nanda Empire and the smaller kingdom of Porus (King Porus ruled the region between the Hydaspes and Acesines rivers), had their own unique troop formations and tactics.

In terms of tactics, Indian armies relied on a combination of their infantry, archers, cavalry, and war elephants to engage and disrupt enemy forces. They would use archers and war elephants to weaken and confuse the enemy.

Image courtesy of Richard Evers.

Infantry

Indian infantry mainly consisted of foot soldiers armed with various weapons, such as bows and arrows, long spears, swords, and shields. They were typically arranged in tight formations, similar to the Greek phalanx, providing a strong defensive line.

Image courtesy of Kevin Krause.

Alexander & His Adversaries

Image courtesy of New Buckenham Historical Wargamers.

Archers

Indian archers were renowned for their skill and the range of their bows, which were often made of bamboo. They used long, powerful bows called *pinaka*, capable of launching arrows over a considerable distance. Archers played a critical role in Indian battle tactics, providing suppressing fire and engaging enemy troops from a distance.

Cavalry

Indian cavalry was less developed compared to their Macedonian counterparts, but they still played a significant role in battles. Indian cavalrymen usually rode on light, fast horses and were armed with spears, swords, and bows, which enabled them to launch hit-and-run attacks on enemy formations.

Elephantry

One of the most distinctive aspects of Indian armies was their use of war elephants. These massive creatures were specially trained and equipped with armor and large tusks. They were used to charge into enemy lines, creating havoc and breaking up infantry formations, and often had archers or javelin throwers on their backs, adding to their offensive capabilities. War elephants posed a significant challenge to Alexander's forces, who were unfamiliar with such adversaries.

Image courtesy of JC Lira.

War Chariots

Indian armies also made use of war chariots, although they were not as prominent as they had been in earlier periods. Chariots called sangramika were typically drawn by horses or sometimes even oxen and were used to transport Leaders, archers, or other troops across the battlefield. Indian Chariots were used by nobles, kings, and warriors as a means to display their status on the battlefield.

Indian war chariots tended to be larger and heavier than Persian scythed chariots, with more elaborate decorations and ornamentation. They often featured ornate carvings, designs, and sometimes even small sculptures or figurines. Curtius (summa virium 8.14.2) tells us that the Indian chariots were four-wheeled and carried six men (two with bucklers, two archers on each side, and two drivers with javelins). The primary purpose of the Indian chariots was to provide a mobile platform for archers to rain down projectiles upon the enemy, while the drivers and men with bucklers provided defensive and offensive support.

Weapons & Armor

When Alexander the Great invaded India in 326 BCE, he encountered various Indian kingdoms and tribes,

Image courtesy of Kevin Krause.

each with its own military forces. The Indian soldiers of that time used a variety of weapons and armor, drawing from a rich military tradition.

Projectiles & Missiles

The Indian longbow, also known as the *dhanush*, was a formidable weapon capable of firing arrows over great distances. The archers used a variety of arrows, including those with broadheads for cutting and those with narrow, hardened tips for penetrating armor.

The *chakram* was a circular metal projectile weapon with a sharp outer edge. As a ranged weapon, it was thrown at foes to cause damage or disrupt formations.

Spears

Different varieties of spears, some designed for thrusting and others for throwing, were used by Indian armies. Some spears had an iron blade in the shape of a leaf, while others had a long, narrow, and pointed tip.

Melee Weapons

Different types of swords were used by Indian soldiers, including the straight, double-edged *khanda* and the curved, single-edged *scimitar*. These swords were primarily employed in close combat for slashing and slicing attacks.

The maces and clubs used by Indian warriors were made of wood, metal, or a combination of both. These weapons were capable of delivering crushing blows that could penetrate armor or stun an opponent.

ARMOR

Some Indian soldiers wore mail or scale armor, which offered protection against slashing and thrusting attacks. This type of armor was created by interlocking metal rings or attaching small metal scales to a leather or fabric base. Indian soldiers also utilized padded or quilted armor made of layers of fabric or leather, which provided a lighter and more flexible form of protection.

Indian warriors wore a variety of helmets, typically fashioned from iron or steel. These helmets provided head protection and occasionally included a face guard or nose guard.

Indian soldiers defended themselves from enemy attacks with shields made of wood, wicker, or metal. Some shields were round or oval, while others were kite-shaped or rectangular.

Indian Army List

INDIA						
Type	Strength	Discipline	Vulner.	Whiff	Ranged	Value
Hvy Infantry	8	0	3	0	bows	41
Med Infantry	6	0	4	0	javelins	24
Skirmish	4	0	5	0	bows or slings	19
Lt. Cavalry	4	2	2	0	bows	23
War Chariot	4	2	1	1	bows	19
Indian Elephantry	4	5	2	-1	bows	38

Leader	Leadership	Tactical Advantage	Value	Country	Inspirations	Limitations
Porus	3	2	17	India	Steadfast	Brash
Sisikottos	2	1	10	India	Resourceful	Aspiring Leader
Assakenos	2	1	10	India	Unbreakable Will	Outnumbered

Alexander & His Adversaries

Era-Specific Rules

Elephantry

When most people think of war elephants in the Roman era, the first thought that comes to mind is probably Hannibal, the Punic Wars and Carthage at war. But, King Pyrrhus of Epirus introduced the Roman Republic to war elephants. The first time Romans fought against war elephants, it was at Heraclea where the Romans were caught off-guard and routed. But later at Asculum, the Romans had prepared several anti-elephantry measures. The Romans battled war elephants with ox-drawn carts with spikes, loud noises and fire. Eventually, Romans themselves would also deploy war elephants in various conflicts in Macedonia, Iberia and even in Britain.

This game does not go into specific anti-elephant measures. An assumption is built into the game that the opposition has some experience and or foreknowledge of elephantry. Despite any measures the opponents may take, war elephants are incredibly dangerous opponents. At the same time, war elephants are very dangerous allies and when they break, things can break very badly.

Image courtesy of Richard Evers.

Elephantry Attributes

Elephant units have 4 Movement points per turn and a discipline of +2. Every one Elephantry figure rolls three dice per point of Strength on the attack. Every one Cavalry figure rolls two dice per figure on the attack. Cavalry is very strong on the initial attack, but can be unreliable in defense.

Elephantry have a number of traits which are different from Infantry and other Troop Types. For example, as described in more detail below, Elephantry are not automatically eliminated from play when they receive more Disruptions than Unit Strength.

When Charging an Enemy Flank, Elephantry roll triple-dice on the Attack. For example, an Elephantry Unit which would normally roll three Attack dice, rolls nine Attack dice.

Elephants are susceptible to Impetuous Charges. If as a result of a Charge, an Elephantry Unit takes ground during the Occupy Phase from a Retreating enemy Unit, the Elephantry Unit must roll a Courage Check. If the check is failed, the Elephantry Unit must immediately move forward and Charge the Retreating enemy Unit. If that unit has already been eliminated, then the Cavalry moves forward to Charge any unit in the next available space. If there are two Units available to Charge, the Attacker chooses which Unit to Charge. The Unit which is not charged may not reply, but the Charged unit may Counter-charge unless they are already committed to a Charge.

If the Courage Check is successful, the Elephantry unit may, at the Leader's discretion, Charge the Retreating enemy Unit.

Image courtesy of New Buckenham Historical Wargamers.

All Impetuous Charges are resolved Sequentially. The Proactive player chooses which of his units to resolve in which order. The Reactive player then resolves any Impetuous Charges.

Impetuous Charges are committed and at the end of the Occupy Phase, but the subsequent Melee is not resolved until the following Melee Phase. That said, Impetuous Chargers will immediately and instantly eliminate any Wavering Unit which the Charger Threatens by Impetuous Charge.

Elephantry is not removed from play when routed. When routed roll on the table below after both sides have moved.

1	**Dazed** - Hold ground unless provoked.
2-3	**Rampage** - Randomly pick a direction and then attack anyone to the front.
4-9	**Rout** - Exit the game causing mayhem on the way out.
0	**Perish** - Remove from play

Dazed

When elephantry is dazed, the Unit turns to a random direction. The Unit does not move or attack and then holds ground. The Unit will remain in this condition for the remainder of the game, unless an ally or enemy attempts to charge or fire upon the unit.

If the unit is charged or attacked at the front, the elephant will attack the unit that attacked. If the unit is charged or attacked in the flank or rear, roll on the table above to find out how the elephants respond.

Rampage

When rampaging, elephantry always rolls 4 dice on the attack, regardless of the unit's current Strength. To begin a rampage, randomly determine a facing for the elephantry.

> ### Random Direction
> When playing with a grid, assign a number to each possible orientation and roll a die to pick an orientation. Reroll any number larger than 6. When playing without a grid, roll a die very close to the unit. The unit turns to face the same direction as the top of the die.

After turning the elephantry to the random facing, move the elephantry a full movement toward the closest unit in the elephantry's forward arc. In a rampage, the elephant does not discern between friend or foe. If more than one unit is at the same distance, randomly pick one of the units to be the target of the rampage.

The target of the Rampage, the unit suffers a 4 die elephantry attack with a Charge bonus. The defender

Image courtesy of Richard Evers.

must Give Ground. If the hex is contested ground, all Charges are canceled and both sides take 2 dice of the attacks. Immediately after the Rampage, the elephant's condition is Dazed.

If no unit is in front of the elephantry, move the elephantry forward a full movement toward the closest possible target. Then change the elephantry's condition from Rampage to Dazed.

Rout

A routed elephantry unit moves 4 Hexes to the front, preferring open to contested, contested to occupied. (Randomly select if two paths are of equal preference.) Any units that are in the elephantry's path suffer 2 disruptions. Any unit in the destination Hex must take 2 disruptions and Give Ground. At the end of the turn, the elephantry remains routed and continues to move in a similar fashion each turn until it exits the battlefield.

Perish

Remove the unit from play.

Indian War Chariot

Several armies described herein used some variation of the war war chariot. A war chariot differs from a war carriage in that a war chariot has only 2 wheels while carriages have 4 wheels. A war chariot differs from say a racing chariot in that it's built for battle and not strictly for speed.

In the earliest days of the ancient era, chariots were an important and popular weapon platform. They were rugged but mostly built for speed accommodating a driver and one or two combatants. These chariots were popular throughout the Levant and the Far East well before the Roman Era, but by the time of Alexander the Great war chariots were superseded by cavalry.

Image courtesy of Richard Evers.

However, into the Roman era the war chariot was still in use by peoples influenced by the war culture and language. The war war chariot differed from the earlier chariots of the Middle East. While chariots in the Levant operated in wide open spaces, the war chariot was built to handle more rugged and wooded terrain found in northwest Europe.

The war war chariot was markedly different from the racing chariots found in Rome and Greece. Rather than a team of large fast horses, many believe that the war chariot relied on 2 stout but sure-footed ponies. There is some debate as to exactly how the war war chariot was used in battle, but many believe it was not a particularly effective weapon platform.

The war chariot was not built for the charge. There were no blades or devices attached to the wheels. It was used as a platform for bows and javelins, but was not particularly effective in this regard because the terrain prevented chariots from assembling in a way to effectively create a real concentration of fire.

We don't presume to answer the question of exactly how the Indian war chariot was used historically, but as a conceit of this particular game, *Alexander & His Adversaries* interprets the war chariot to be a sort of armored personnel carrier or APC. As an APC, the war chariot is no tank, but neither is it without some offensive and defensive capabilities.

Mounting and Dismounting

At any point during its movement, a war chariot can take on and discharge passengers. There are limits to how many Troops a chariot can transport and there are some requirements for the passengers to mount and dismount.

The chariot may only transport a Unit with 6 Troops or less. Disrupted Troops count toward this limit but Casualties do not. When the transported Unit's Strength is 4 or less, the chariot retains its offensive capabilities when transporting friendlies. But when transporting 5 Troops or more, the chariot has no offensive capabilities.

A Unit can only mount a chariot if it has not yet moved this turn. A unit can mount the chariot during the chariot's move but mounting costs half of the transported unit's movement to mount. Similarly, the transported unit can dismount during the chariot's movement, but dismounting costs half of the transported unit's movement.

The unit can mount the chariot any Adjacent Hex at any point of the chariot's movement. The unit can dismount into any Adjacent Hex at any point of the chariot's movement. A Unit may not charge on the same turn as it dismounts. However the dismounted unit can face in any direction.

- **Speed** - A chariot unit has 4 Movement points.
- **Maneuver** - Once per Movement Phase, a chariot can turn up to 180 degrees without paying a point for Maneuver.
- **Ranged Weapons** - A war chariot may fire bows during the Movement Phase. If it does so, that unit may not fire again in the Ranged Phase.

Image courtesy of New Buckenham Historical Wargamers.

Appendix I: Blood of Ancients

Summary Rules

While this summary does not make any great effort to provide a detailed explanation of the terms or concepts that are fundamental to the game, it does provide some context and the framework of **Blood of Ancients**.

Order of Play

Initiative

Both sides roll a die. The higher result is the Proactive side. The lower result is the Reactive side.

In the event of a tie, the side that won the previous initiative loses.

Image courtesy of Richard Evers.

Command

First the Proactive side then the Reactive side moves all leaders. Leaders may move up to 4 Hexes in any direction without regard for maneuver. By default, all Leaders have a Command Radius of 4 Hexes.

Movement

First the Proactive side then the Reactive side moves units. Each unit is moved one at a time resolving one unit's movement before starting another unit's movement. In general infantry and foot troops have 3 points of movement and cavalry has 5 points of movement

In the Movement Phase, charges and counter-charges are declared.

Ranged

Resolve all Ranged weapon fire. All fire is considered simultaneous. Thrown weapons have a range of 1 Hex. Projectiles such as slings and arrows have a range of 4 Hexes.

Rolling Attacks

For each point of Strength the Unit currently has, roll one attack die. Each die that is equal to or lower than Vulnerability is a hit.

> ### About Charges
>
> The heart of Blood of Ancients is the charge system. When an attacker Charges, they get a bonus to their attack rolls. However, the opponent can Counter-Charge to get the same bonus. The actual melee occurs in Contested Ground between two charging units, or at the spot where a defender waits. The victor of the melee assumes the Contested Ground.

When determining hits, the Attacker adds Whiff to his dice. Because a low roll is desirable, the higher a Unit's Whiff the more likely the attack will be a miss. That means that a negative Whiff is a bonus to hit.

Rule of Aces: regardless of circumstances, Vulnerability or Whiff, a 1 is always a hit and a 10 is always a miss.

Courage

Units entering Melee must make a Courage Check. Fresh Units automatically pass their Courage Checks. Units that fail their Courage Check do not Charge, and they must immediately roll a Morale Check.

Image courtesy of Richard Evers.

Image courtesy of Richard Evers.

Units which fail a Morale Check will Retreat. Unit's that pass this Morale Check hold ground with benefit of Bracing, but the opponent occupies the Contested Ground at this time.

A Courage Check is not required when charging a defender's Rear or Flank.

Melee

Resolve Melees in any order. All combat is considered simultaneous.

Morale

Immediately remove all Routed Infantry Units from play. Remove any Wavering Infantry Units from play if they are Adjacent to an enemy Unit. Any Wavering or Routed Cavalry Units immediately Retreat. Any Fatigued Unit which is Threatened must roll a Morale Check or Retreat.

All Retreats, see also next Phase, are simultaneous.

Retreat

Retreating causes great disorder in the Unit, so place two additional Disruptions on the retreating Unit-- which can cause a unit to Rout rather than Retreat.

When a Unit retreats, place a Disruption on any Adjacent allied Units. Retreats have a number of requirements and limitations which are described fully in the Glossary.

It's possible for both sides to Retreat. A Unit is only required to pass one Morale Check per Morale Phase and multiple adjacent Friendly Retreats do not confer only one penalty for a total of 2 Disruptions.

Is your unit Fresh or Wavering?

Fresh Units have no Disruptions. Fatigued Units have one or more Disruptions. Wavering Units have Disruptions equal to their Strength. Routed Units have more Disruptions than their current Strength.

ALEXANDER & HIS ADVERSARIES

> ### More Details
>
> There are certain penalties and limitations associated with Intermingling which are described in the Glossary of Blood of Ancients. In fact, any of the terms you see capitalized have reserved meanings that you can find defined in the Glossary.

Occupy

Melee victors occupy Contested Ground. Only those Units that charged may occupy the Contested Ground and those Units may only occupy the Contested Ground that they charged.

When advancing into the Contested Ground causes the occupying Unit to come into contact with Wavering Units, Wavering Units are removed from play.

If several friendly Units attempt to occupy a single hex in the Occupy Phase, the Units are Intermingled.

Recovery

During the Recovery Phase, all Units not Adjacent with the enemy have the option of removing all Disruptions. The recovering Unit can form up Facing any direction. Each time a Unit recovers, remove one figure from the Unit as a Casualty. A Casualty cannot be recovered.

> ### A Good General knows ...
>
> A wise general knows when to recover and when to press. Recovery is always optional.

Image courtesy of New Buckenham Historical Wargamers.

Appendix II: Personalizing Fit To Your Collection

Blood of Ancients does not rely upon a particular scale of model. The game assumes a hex grid, but does not *require* a hex grid. To facilitate scaling the game to different sized figures the word "Hex" (capitalized) as a reserved game term. As a reserved term, a Hex is not a shape.

Instead, a Hex is defined as a linear unit of measurement, like an inch or centimeter. Unlike a centimeter or an inch, a Hex is not a fix unit of measure, but rather the Hex is a scalable unit of measurement.

Although a Hex can vary in size, this does not mean that in any given game, the size of a Hex will change.

Image courtesy of Kevin Krause.

In any given game, all figures should be roughly the same scale and the size of a Hex is constant. The game master decides the size of a Hex before the game starts.

Determine your ideal Hex size by putting a unit of models on the table. Use dice or tokens to mark a square around the unit. The square should be as small as possible, but the entire unit should be within the square. Even if your unit is rectangular, the area around the unit must be square. Then measure one side of the square. Round that to a gamer friendly number and the result is the size of a Hex.

Of course, there are an infinite number of choices, but we've found that 4 inches or 10 cm per hex is pretty common and works best with a variety of scales. Ranges and most other measurements in Blood of Ancients are given in Hexes, so once you know the size of a Hex, you should be able to translate most of the game easily to a game without a grid.

We track Strength on the tabletop with models and tokens. Strength is a numeric representation of how many dice a unit rolls on the attack. It's also the base Discipline for a unit. A Unit's current Strength is Strength minus the number of Disruptions on the Unit.

Blood of Ancients basic rules generally assume that 1 point of Strength equals 1 model on the table. But, depending on how you define a Hex, you may need to redefine the ratio of Strength to models. For example, you might want each model to represent 2 points of Strength. When a model represents more than 1

Image courtesy of Kevin Krause.

point of Strength, you will need to use tokens to track Casualties and Disruptions.

12 divides nicely by the following numbers: 1, 2, 3, 4, 6 and 12. If you diverge from one of these factors, you really are better off using tokens beside the models to track the Strength of units during play. But if we take a look at those ratios, it's easy to see which factors are easiest to play with. In the table below, a "Weakness" is a token placed beside the unit to represent a reduction in Strength by 1 point. A "Strength" on the table below, is a token placed beside the unit to represent a bonus of 1 to Strength.

Factor Unit Size	1	2	3	4	6	12
4	4 models	2 models	1 model + 1 Weakness	1 model	1 model + 2 Weakness	1 model + 8 Weakness
6	6 models	3 models	2 models	2 models + 2 Strength	1 model	1 model + 6 Weakness
8	8 models	4 models	2 models + 2 Weakness	2 models	1 model + 2 Strength	1 model + 4 Weakness
10	10 models	5 models	3 models + 1 Weakness	2 models + 2 Strength	2 models + 2 Weakness	1 model + 2 Weakness
12	12 models	6 models	4 models	3 models	2 models	1 model

Of course any of the ratios in that table will work, but the two best ratios to play with are probably 1:1 and 1:2.

Image courtesy of Richard Evers.

Alexander & His Adversaries

Image courtesy of New Buckenham Historical Wargamers.

Appendix III: Playing Without A Hex Grid

Movement

In *Blood of Ancients*, movement and ranges are given in Hexes. When you're playing without a grid that means that each point of Movement allows you to move up to 1 Hex forward. Movement does not allow the unit to turn (even a little) or change facing. While a unit may not turn, the unit can sideslip up to 45 degrees to the right or to the left without turning.

Any movement is Movement Point

Any movement from zero to one Hex costs one point of Movement. For example, if the Hex is 4 inches and the unit moves forward one-eighth of an inch that costs 1 point of movement. If the Hex is 10 centimeters and a unit moves forward 1 millimeter, 8 millimeters or 10 centimeters, that costs one point of Movement.

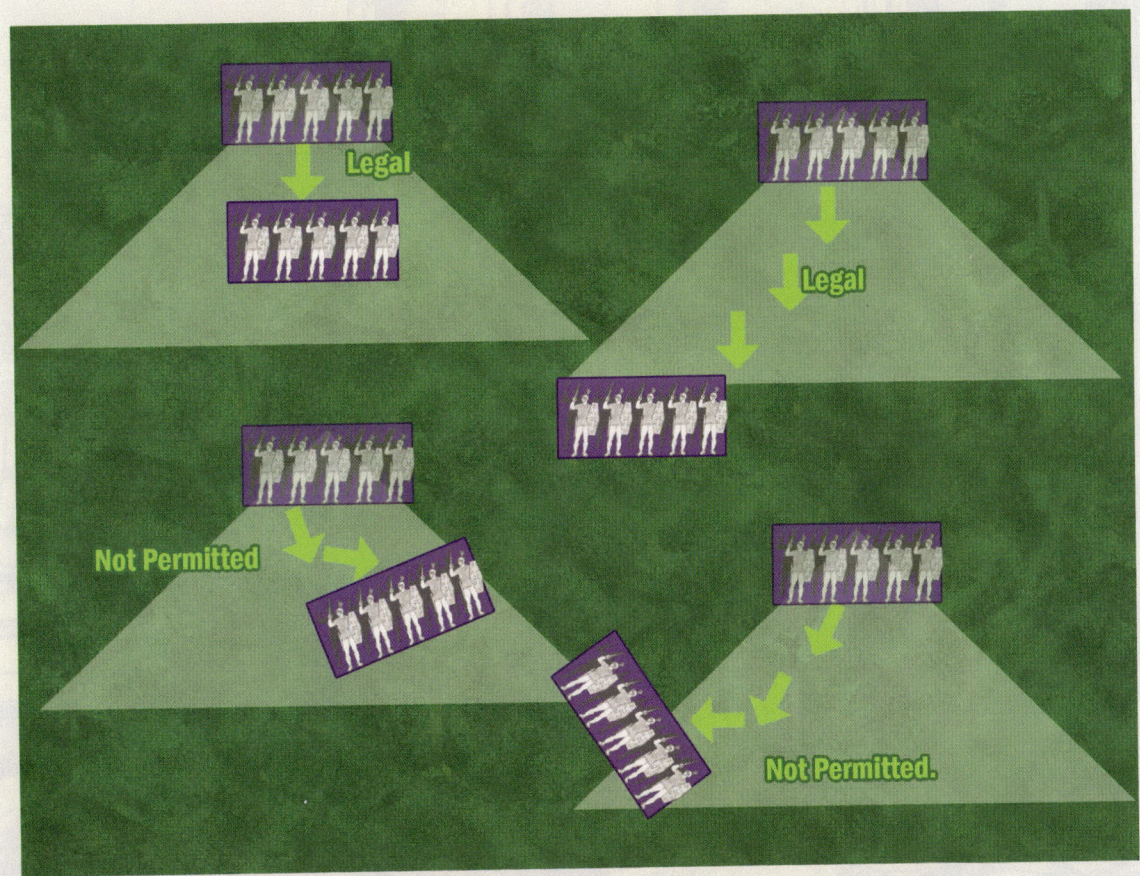

Maneuver

Maneuver is how players change the facing of a unit. Each Maneuver allows a unit to turn its facing up to 45 degrees. Each Maneuver costs 1 Hex of Movement. A unit can move and maneuver in the same turn and the same phase, but not at the same time.

Any change of facing (called a Maneuver) costs 1 Hex. A turn of 1 degree costs 1 Hex. A turn of 45 degrees costs 1 Hex. A turn of 90 degrees costs 2 Hexes.

A unit can move before maneuvering. If the unit has a Hex remaining after making a Maneuver, the unit can continue to Move.

Breaking Formation

The Break Formation rule only applies to those games where there is no hex-grid in play. Breaking Formation is useful when the move and maneuver rules are too cumbersome to just move your unit a few inches this way or that. Breaking Formation allows a player to move a Unit in any direction and end up with any facing. The catch is that when a unit breaks formation, the unit can only move half of its total movement.

For example, let's look at a unit that has 3 Hexes of movement in a game that has established a 4-inch Hex. If that unit moved straight forward it could move 12 inches. However that same unit can Break Formation and move 6 inches in any direction and end up facing in any direction.

Image courtesy of Kevin Krause.

Alexander & His Adversaries

A unit that breaks formation cannot Charge, but it can move to attack without a charge bonus.

Charging Without Hexes

The rules for charging without a hex grid are the same as the rules for playing with a hex grid. The only thing that becomes a bit more nebulous is Contested Ground. As described in *Blood of Ancients*, melees occur on Contested Ground. Contested Ground can be either the location that a defending occupies, or an area between a charging unit and a counter-charging unit.

Defender Receives

If the defender braces to receive the charge and elects not to counter-charge, then the Contested Ground is the defender's current location. The attacker automatically advances to a spot about 1" (25 mm) from the defender. Don't worry about squaring up the units or "conforming" as is the practice in some games. The resulting melee will decide who conforms and which unit flees. So for now, just move the units so that they are nearly in contact with each other.

Image courtesy of Richard Evers.

Counter Charge

If the defender counter-charges, the charging Unit moves within one Hex of the enemy. For example if your Hex is 4 inches then move within 3 inches or so of the enemy unit. As above there is no need to conform or align the units. The space between the units is the Contested Ground and after the melee one of those two units will (in all likelihood) occupy the Contested Ground. We indicate this with a charge token and or by moving a few models from both the charger and the counter-charging Unit into the Contested Ground.

Gridless Arc of Fire

When playing without a grid, the arc of fire proceeds out at a 45 degree angle from straight forward.

Occupy Phase

The Occupy Phase is when we resolve the facing of Units that were Meleeing over Contested Ground.

The Gridless Retreat

A Unit which fails a Morale Check must Retreat. Retreating causes great disorder in the Unit, so place 2 additional Disruptions on the retreating Unit. Furthermore, a retreat is very disheartening to allied Troops. When a Unit retreats, place a Disruption on any allied Units that are less than 1 Hex away.

These Disruptions take effect immediately so the Retreating player may wish to take care which order he rolls his Morale check.

In addition to suffering extra Disruptions, the retreating Unit must move a full move away from the enemy from its current position. The retreating Unit does not pay for any maneuver and completes the Retreat move facing any direction the retreating player desires.

A retreating Unit must increase its distance from enemy Units during the entirety of the Retreat. If anytime during the Retreat, the retreating Unit cannot increase its proximity to the enemy, then the Retreating Unit is removed from play.

The exception to this rule is when the retreating Unit is retreating toward a friendly Unit that lies between the enemy and the retreating Unit.

A Unit may never Retreat into or through any Contested Ground.

Both Sides Retreat

It's possible that all Units from both sides Retreat. When this happens neither side occupies the Contested Ground.

One Side Retreats

When only one side Retreats, then the Troops that did not Retreat occupy the Contested Ground. Only those Units that charged may occupy the Contested Ground and those Units may only occupy the hex that they actually charged. The occupying Unit may turn up to 45 degrees after pushing forward into the contested ground.

When advancing into the Contested Ground causes the occupying Unit to come within 1 quarter Hex of enemy Wavering Units, Wavering Units are removed from play. If several friendly Units attempt to occupy a single hex in the Occupy Phase, the Units are Intermingled. There are certain penalties and limitations associated with Intermingling which are described in the Blood of Ancients Glossary.

Neither Side Retreats

If neither side Retreats, place one Disruption on each Unit. Both units turn (in place) up to 45 degrees to face each other. Both units should turn to the same amount as they become enmeshed in melee. Furthermore both units should move in place to left or right up to one quarter of a Hex. The purpose of this side-step is to bring the center of their front rank of troops into alignment.

In this case, neither side occupies the Contest Ground. If these Disruptions cause one unit or another to Rout, the Routed Unit is removed from play, but neither side occupies the Contested Ground.

Image courtesy of Richard Evers.

Image courtesy of New Buckenham Historical Wargamers.

Printed in Great Britain
by Amazon